Signals and Communication Technology

This series is devoted to fundamentals and applications of modern methods of signal processing and cutting-edge communication technologies. The main topics are information and signal theory, acoustical signal processing, image processing and multimedia systems, mobile and wireless communications, and computer and communication networks. Volumes in the series address researchers in academia and industrial R&D departments. The series is application-oriented. The level of presentation of each individual volume, however, depends on the subject and can range from practical to scientific.

Indexing: All books in "Signals and Communication Technology" are indexed by Scopus and zbMATH

For general information about this book series, comments or suggestions, please contact Mary James at mary.james@springer.com or Ramesh Nath Premnath at ramesh.premnath@springer.com.

Chua-Chin Wang • Arumugam Nallanathan
Editors

Proceedings of the 5th International Conference on Signal Processing and Information Communications

Springer

Editors
Chua-Chin Wang
National Sun Yat-sen University
Kaohsiung, Taiwan

Arumugam Nallanathan
School of EE and Computer Science
Queen Mary University of London
London, UK

ISSN 1860-4862 ISSN 1860-4870 (electronic)
Signals and Communication Technology
ISBN 978-3-031-13180-6 ISBN 978-3-031-13181-3 (eBook)
https://doi.org/10.1007/978-3-031-13181-3

This Springer imprint is published by the registered company Springer Nature Switzerland AG
The registered company address is: Gewerbestrasse 11, 6330 Cham, Switzerland

Preface

It is truly my pleasure to welcome all of you to join the 2022 5th International Conference on Signal Processing and Information Communications (ICSPIC 2022). Starting out as a workshop for the first time in 2018 to promote research exchange in the field of information processing, AI, and communication, this event has grown into a medium-scale virtual symposium this year with a high-quality technical program and an international forum. Not only ICSPIC 2022 publicizes many state-of-the-art papers in the mentioned research areas recommended by a strong International Program Committee organized by members around the globe but it also holds three plenary talks presented by world-renowned researchers. The success of ICSPIC 2022 certainly paves the path for a better event in the future.

Due to the long-term impact of COVID-19, we were unable to meet in Paris, France, as planned. ICSPIC 2022 used ZOOM as the supportive platform to allow all the participants around the world to gather online at the same time. To make sure that all the participants can use ZOOM smoothly, the organizing committee arranged a test session to teach them the basic functions of ZOOM. In the following days of the conference, there were keynote speeches, invited speeches, and technical sessions. Though the authors and speakers couldn't communicate face to face, the passion for involvement wasn't affected.

This volume of proceedings contains 13 papers. From the proceedings, readers can avail the most cutting-edge knowledge about signal processing and information communication from all around the world. And each paper was presented at the conference. Topics include digital signal and image processing, modern information technology, and management. Each selected paper went through a double-blinded peer review. Our warm gratitude firstly goes to the committee members. The paper control work was really a hard task, which cost a lot of time and energy. The committee members had to choose papers from all the excellent submitted research works based on their professional achievements and profound knowledge. Without their contribution, the conference agenda would not have materialized, and the proceedings couldn't have been generated.

We hope that ICSPIC will become better and better, and it can become a leading conference in this field with your high supports one day.

Kaohsiung, Taiwan Chua-Chin Wang

Contents

Part I Digital Signal and Image Processing

A Deep Learning-Based Approach for Camera Switching in Amateur Ice Hockey Game Broadcasting 3
Hamid Reza Tohidypour, Yixiao Wang, Mohsen Gholami, Megha Kalia, Kexin Wen, Lawrence Li, Mahsa T. Pourazad, and Panos Nasiopoulos

Performance of LDA Versus PCA in Nonideal Lighting Environment 13
Hikmat Darwish

The Effect of the Audio Signals on the Brain for EEG-Based Attentiveness Recognition 29
Jiaming Cai, Pinzhen Chen, Yunyang Yuan, and Zecheng Li

EMG-Based View Controller Using VR Applications 39
Jingcheng Zhao, Anjun Zhang, Tianao Zeng, Hao Cai, and Yanlong Wang

Comparison and Application of Two Face Detection Algorithms 53
Xiaoxi Wei and Qian Yin

Research on Sound Field Evaluation of Urban Walking Street Through Sound Environment Perception 63
Wei Lin and Yi-Ming Wu

Part II Modern Information Technology and Management

Implementation of Machine and Deep Learning Algorithms to Detect Gaps in a Sheet Metal Industry: A Case Study 77
Ahm Shamsuzzoha, Timo Kankaanpaa, Huy Nguyen, and Hoang Nguyen

Modelling the Service Process at the Cash Register Using Generating Functions 89
Andreas Ahrens, Detlef Hartleb, and Jelena Zăšcerinska

Adaptive, Automatic and Non-invasive Cultural Heritage Preventive Conservation Framework Based on Visual Information Crowdsourcing 103
Miguel Antonio Barbero-Álvarez, Juan Antonio Rodrigo, and José Manuel Menéndez

An Inverted-ITL Algorithm for Mining Partial Periodic-Frequent Patterns 117
Ye-In Chang, Xin-Long Chen, and Sheng-Hsin Chiang

A Viewpoints-Based Analysis of Enterprise Architecture Debt 133
Ather Maqsood, Peter Alexander, Horst Lichter, and Sansiri Tanachutiwat

Detection and Prevention of ARP Cache Poisoning in Advanced Persistent Threats Using Multiphase Validation and Firewall 155
Muaadh Nasr Al-Mwald, Norziana Jamil, Zul Azri Ibrahim, Zaihisma Che Cob, and Fiza Abdul Rahim

Permeability and Inertia Coefficient of Wire-Mesh Porous Media Estimated by a Darcy-Forchheimer Equation 171
Suradech Sinjapo, Panuwat Chanmak, and Bundit Krittacom

Index 185

Part I
Digital Signal and Image Processing

A Deep Learning-Based Approach for Camera Switching in Amateur Ice Hockey Game Broadcasting

Hamid Reza Tohidypour, Yixiao Wang, Mohsen Gholami, Megha Kalia, Kexin Wen, Lawrence Li, Mahsa T. Pourazad, and Panos Nasiopoulos

Abstract Switching the camera views in ice hockey plays an important role in television broadcasting. The traditional approach is to employ a broadcast director and professional crew responsible for making decisions about switching camera views, which is not an affordable option for amateur games. In this study we propose an automatic switching scheme of camera views that is based on a deep learning-trained model which detects important objects such as players, puck, and net. Our solution uses a Faster-RCNN object detection network which was trained using a dataset of 1000 high-definition (HD)-labeled frames of hockey games. Our Faster-RCNN achieves the average precision (AP) of 0.90, 0.88, and 0.61 for players, net, and puck, respectively, using a vgg-16 pretrained model. Our camera switching approach uses the confidence values of the detected objects to predict the best camera view for the current moment. Performance evaluations showed that our method achieved accuracy of 75% for choosing the most important camera view in real time.

Keywords Multi-camera switching · Object detection; Faster-RCNN · Convolutional neural network · Video and camera tracking

H. R. Tohidypour (✉) · Y. Wang · M. Gholami · M. Kalia · K. Wen · L. Li · P. Nasiopoulos
Department of Electrical & Computer Engineering, University of British Columbia, Vancouver, BC, Canada
e-mail: htohidyp@ece.ubc.ca; yixiaow@ece.ubc.ca; mgholami@ece.ubc.ca; mkalia@ece.ubc.ca; kwen04@ece.ubc.ca; panos@ece.ubc.ca

M. T. Pourazad
Department of Electrical & Computer Engineering, University of British Columbia, Vancouver, BC, Canada

TELUS Communications Inc., Vancouver, BC, Canada
e-mail: pourazad@ece.ubc.ca

C.-C. Wang, A. Nallanathan (eds.), *Proceedings of the 5th International Conference on Signal Processing and Information Communications*, Signals and Communication Technology, https://doi.org/10.1007/978-3-031-13181-3_1

1 Introduction

It is common practice in broadcasting live sports to utilize many cameras and switch between them, depending on how the game develops. This switching involves professional operators and expensive specialized equipment. Besides the expensive equipment, professional coverage involves highly skilled individuals such as camera operators and a director responsible for supervising and deciding the overall operation.

Unfortunately, such an expense is prohibitive when it comes to broadcasting amateur community or school sports. In this case, despite the fact that more than one camera may be used, real-time coverage involves only the main view, without offering the option of watching the view that better covers crucial moments during the game.

As a result, this monotonous coverage of regional sports may potentially hinder the viewership and be detrimental in the progress of school and amateur games. Thus, there is a need for a cost-effective, fully automated camera view switching system, which analyzes the importance of the scene covered by each camera and then switches the view in a manner that is pleasant to the viewer.

To the best of our knowledge, there is only one existing work for designing automatic camera switching for ice hockey which is presented in [1]. This method performs an automatic camera selection using the hidden Markov model to create personalized video programs for users that are more interested in the performance or positions of the players from different perspectives than the game itself. In that respect, this method was player-centered rather than puck-centered or play-centered. However, our task is to design an automatic play-centered camera switching approach for amateur ice hockey games.

To this end, in this chapter, we propose a play-centered solution that is based on deep learning, namely, the Faster-RCNN architecture [2], to optimize view switching in regional ice hockey games. Our deep learning-based object recognition network receives video feeds from the two primary camera views of ice hockey and detects the players, net, and the puck in real time with very good precision. Then, based on the predicted confidence values for the different objects, our algorithm decides which camera view should be broadcasted.

The rest of this chapter is organized as follows. Section II presents our approach and explains the dataset selection and labeling of our dataset. Section III presents the performance evaluation of our method and discusses the results. Finally, Section IV concludes our chapter.

2 Our Approach

To keep viewers of an amateur ice hockey event engaged, there are primarily two types of views that are important; first, the side view that shows the arena (please see left image in Fig. 1) and gives a wide view of the field, and second, the goalie views (please see right image in Fig. 1) that show a closer view of the nets. It is common practice to use these cameras in amateur community or school ice hockey games. Since, most of the action is taking place away from the net, the primary view is the arena (side) view, while the goalie views are the secondary views. Thus, the natural question that arises here is when to show the goalie view, in other words, what are the criteria that will lead to switching from showing the side view to a goalie view. To address this challenge, we first propose the use of a deep learning-based object recognition approach that receives the video feeds from all the views and detects the players, puck, and the nets. Then, we use the weighted sum of the confidence values of the detected objects to decide which view to broadcast. Figure 1 shows our proposed scheme. Details regarding the dataset that we used to train our network and the criteria we introduced for switching camera views are presented in the following subsections.

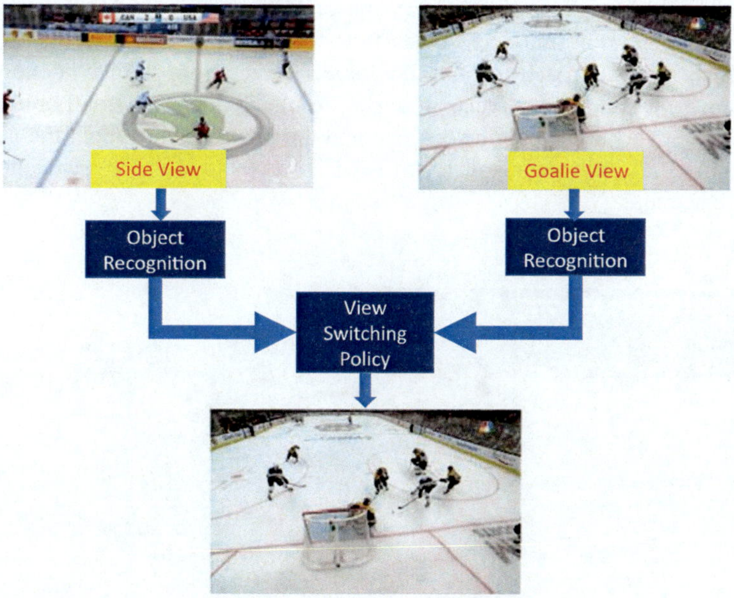

Fig. 1 Our proposed scheme for automatic camera view switching

2.1 Data Collection

In order to build a comprehensive dataset for our application, we downloaded several hockey videos of the resolution of 1920 × 1080 from YouTube [3]. The reason for turning to YouTube and not using amateur content was that we did not have access to the latter due to covid-19 restrictions, which did not allow community games to take place and could not find recorded content of previously played games. From those videos, 1000 representative frames were selected for the training-validation phase, skipping redundant frames and considering only frames with significantly different content to avoid overfitting, preferably including the puck and of high visual quality – avoided blurry, fast-moving puck frames. As already mentioned, in this study the objects of interest were the players, net, and the puck, while the referees and audience were excluded. The location of these three types of objects can be used to determine the best camera view for the current situation. An example of a labeled training frame from the side view is shown in Fig. 2a. Figure 2b shows a different example from the goalie view. For the test phase, we used four ice hockey video streams from YouTube, which were very different from the training videos [3]. These videos had the same resolution with the training videos.

2.2 Our Deep Learning Network

We chose the Faster-RCNN architecture [2] as our deep learning-based classification and object recognition network. The main reason for this choice is that Faster-RCNN is proven to be more accurate and much faster compared to its predecessors [4–6], making it an ideal approach for real-time object detection of the ice hockey fields [2]. Moreover, it also showed very promising results in detecting small objects. We trained this network to detect players, net, and the puck. Details about the network configuration and the training platform used are explained in the evaluation and discussion section.

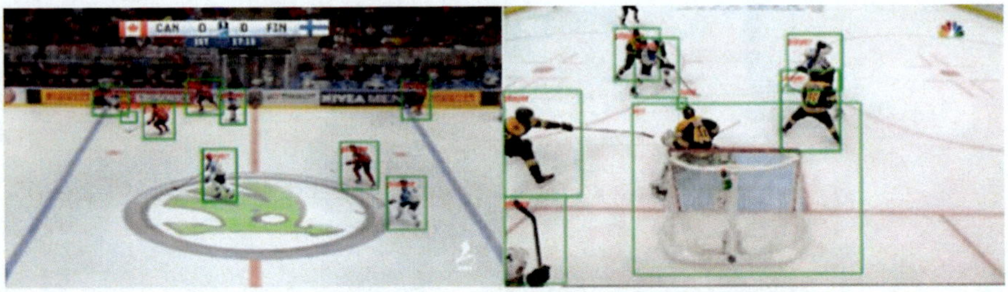

a) b)

Fig. 2 Examples of labeled frames from our dataset (**a**) from the side view and (**b**) from the goalie view

2.3 *Our Camera Switching Approach*

The first task of our scheme for switching camera views is to receive the detection information of the objects of interest, i.e., the players, net, and puck, that comes out of our Faster-RCNN object recognition model. Then, our algorithm considers the position and confidence level of detection of all the objects, as each one has different roles to play in determining the best camera view for the current moment of the game. It is important to note that designing our algorithm to be biased toward the importance of objects to the fans, will allow our solution to be focused on the action. Driven by professional game coverage, we assume that the most important object/ event in hockey broadcasting involves the puck, as the audience tries to follow its location when watching a hockey game. Following the above observation and the outcome of many trials asking subjects to validate the validity of our switching scheme, we assigned a weight to the confidence values predicted for each object type according to its importance: 20 for the puck, 1 for the net, and 1 for the player. More precisely, the confidence of each detected object in the current camera view is weighted according to its object type, and the weighted values are summed up to calculate the score for the current camera view. Please note that our method only considers objects with confidence values greater than 20%. In addition, we decided to add 10 to the weighted score calculated for the goalie view if the puck is present in that camera view.

Figure 3 shows the block diagram of our proposed camera switching scheme. To prevent any high-frequency camera switching, we built in a small delay of ten-frame duration (one-third of a second) before switching again after the last camera view change.

3 Evaluation and Discussion

For training, we used a PyTorch implementation of Faster-RCNN [7]. The fully connected layer of the model was changed to detect the three classes required for our application. Ninety percent of the training-validation dataset was randomly selected as the training dataset and the remaining 10% was considered for the validation phase. Horizontal data augmentation was used to augment the dataset for the training phase. For the training phase, we aimed to achieve the best performance by testing different combinations of the network configurations for this phase. To this end, two different pretrained models, namely, VGG16 and ResNet-101, were used as the backbones for the Faster-RCNN. Four batch sizes, namely, 1, 6, 12, and 24, were tried. Three different learning rates were used to achieve the best training performance: 0.0001, 0.001, and 0.01. We trained our Faster-RCNN using the Nvidia V100 Volta GPU, with 32 GB of HBM2 memory available on a state-of-the-art

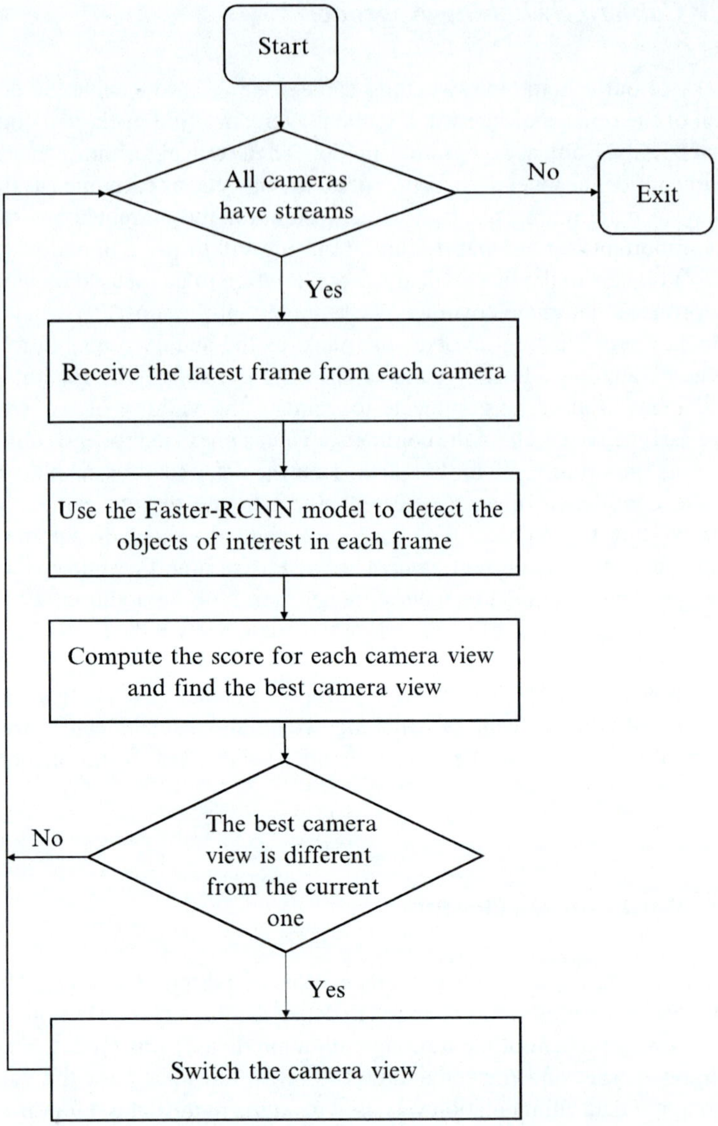

Fig. 3 Our proposed camera switching scheme

advanced research computing network [8]. Tables 1 and 2 show the average precision (AP) of the player, puck, and net classes that were achieved for the validation frames for each of the training settings. The batch sizes (bs) and learning rates (lr) used for each training setting are also reported in these tables. As can be seen in Table 1, the best AP values that were obtained by Faster-RCNN with VGG-16 backbone were 0.616, 0.876, and 0.9 for the puck, player, and net, respectively. The

Table 1 Detection results of Faster-RCNN using different settings with VGG-16 pretrained model

bs	lr	epoch	AP puck	AP player	AP net	mAP
1	0.001	20	0.205	0.821	0.518	0.518
6	0.001	20	0.565	0.876	0.896	0.779
12	0.001	20	0.382	0.876	0.875	0.711
24	0.01	20	0.52	0.88	0.875	0.758
12	0.01	20	**0.616**	**0.876**	**0.9**	**0.789**
12	0.0001	20	0.205	0.821	0.518	0.518

Table 2 Detection results of Faster-RCNN using different settings with ResNet-101 pretrained model

bs	lr	epoch	AP puck	AP player	AP net	mAP
1	0.001	15	0.489	0.865	0.844	0.732
6	0.001	20	0.565	0.876	0.896	0.779
12	0.001	20	0.382	0.876	0.875	0.711
24	0.01	20	0.489	0.865	0.844	0.732
12	0.01	20	**0.587**	**0.878**	**0.87**	**0.778**
12	0.0001	20	0.347	0.86	0.7	0.636

mean average precision (mAP) was 0.789. This performance was achieved using the learning rate of lr = 0.01 and batch size bs = 12. According to Table 2, the best AP obtained with the ResNet-101 pretrained model was 0.587, 878, and 0.778 for puck, player, and net, respectively. The mAP for this case was 0.778. This clearly shows that VGG-16 pretrained model, batch size bs = 12, and learning rate lr = 0.01 achieved the best performance among all the training settings examined in our study. Therefore, for the test phase, we used this model and we call it our model, hereafter.

In order to evaluate the performance of the trained model, we examined the trained model on the test videos with unseen frames. Our results showed that for most of the frames, our model detected the players, net, and the puck correctly. Figure 4 shows the predicted objects and the probability values assigned to the bounding boxes for four successful examples. However, there were some false positives and false negatives for the puck during the test phase. Some examples are shown in Fig. 5. Our results showed that our model detected the puck correctly when it was fully visible to the camera (see Fig. 5a). Since puck is a very small object compared to the players and the net, this contributes to the low AP of the puck. More precisely, our model could not detect the puck when it was not fully visible or blurry to the camera (Fig. 5b). Figure 5a shows an example of false positive for the puck, a case we observed only for few frames. In this rare case, the toe of the hockey stick was detected as a puck since its color was the same as the puck and the color of the hockey stick's blade was the same as the ice hockey field. As can be seen, in this case, two objects were detected as the puck. Thus, we decided to only consider the object with the highest predicted confidence. This approach significantly improved our accuracy. This issue may be resolved by adding frames with a similar scenario to the original training frames. Finally, we used the information of the objects detected

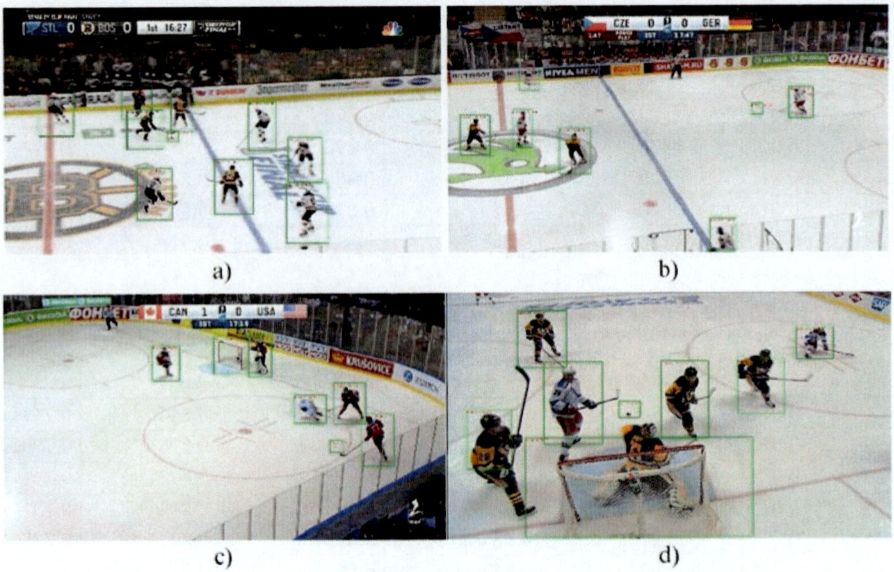

Fig. 4 Example frames from the test set: (**a**, **b**) players and the puck that were detected correctly by our model; (**c**, **d**) players, the puck, and one of the nets that were detected correctly by our model

Fig. 5 Example frames from the test set: (**a**) players, the net, and the puck that were detected correctly by our model as well as a false positive puck; (**b**) players and the net that were detected correctly by our model as well as a non-visible puck (false negative)

by our model for our automatic camera view switching approach to detect the instances for which the view switching was needed. Our results show an accuracy of 75% for our camera switching method in real-time. Considering the fact that only 1000 frames were used for the training and validation phases, our camera switching approach achieved a great performance.

Acknowledgments This work was supported in part by the Natural Sciences and Engineering Research Council of Canada (NSERC – PG 11R12450) and TELUS (PG 11R10321). This research was enabled in part by support provided by WestGrid (www.westgrid.ca) and Compute Canada (www.computecanada.ca).

References

1. L. Wu, Multi-view hockey tracking with trajectory smoothing and camera selection. Thesis, University of British Columbia. Retrieved from https://open.library.ubc.ca/collections/ubctheses/24/items/1.0051270 (2008)
2. S. Ren, K. He, R. Girshick, J. Sun, Faster r-cnn. Towards real-time object detection with region proposal networks. Adv. Neural Inform. Proc. Syst., 91–99 (2015)
3. 2019 IIHF Ice Hockey World Championship, IHF Worlds 2021, YouTube, https://www.youtube.com/c/IIHFWorlds/videos. Last accessed 2021/11/17
4. R. Girshick, J. Donahue, T. Darrell, J. Malik, Rich feature hierarchies for accurate object detection and semantic segmentation, in *Proc. IEEE conference on computer vision and pattern recognition (ICCV)*, (Columbus, 2014), pp. 580–587
5. R. Girshick, Fast r-cnn, in *Proc. IEEE international conference on computer vision*, (2015), pp. 1440–1448
6. K. He, X. Zhang, S. Ren, J. Sun, Spatial pyramid pooling in deep convolutional networks for visual recognition. IEEE Trans. Pattern Anal. Mach. Intell. **37**(9), 1904–1916 (2015)
7. S. Ren, Faster-RCNN, GitHub repository, https://github.com/ShaoqingRen/faster_rcnn. Last accessed 2021/11/17
8. Compute Canada state-of-the-art advanced research computing network. https://www.computecanada.ca. Last accessed 2021/11/17

Performance of LDA Versus PCA in Nonideal Lighting Environment

Hikmat Darwish

Abstract Automatic face recognition (AFR) is used to identify people by processing their photos or snapshots of faces, either in off-line or real-time manners, respectively. However, classical face recognition techniques have been reported to suffer from substantial degradation in performance when person image is subjected to nonideal lighting or some types of occlusion. In real life we may well encounter a certain type of nonideal lighting such as side-shadowing of the face, where substantial part of the face can be totally occluded or masked. In this work, we examine and evaluate the performance of two famous statistical approaches for AFR namely PCA and LDA in terms of face recognition rate (FRR), when both are operating on particular ill-illuminated image exemplified by side-shadowing occlusion with addition of "salt-and-pepper" noise, which is often the encountered case. The two suggested AFR techniques are the well-reputed principal component analysis (PCA) and linear discriminate analysis (LDA). A computer simulation has been executed testing both PCA and LDA and the simulation outcomes indicate much better performance of LDA over PCA in terms of FRR for this particular type of image occlusion.

Keywords Automatic face recognition (AFR) · Principle component analysis (PCA) · Linear discriminate analysis (LDA) · Covariance matrix · Eigen vector · Feature vector · Minimum Euclidian distance · Rate of recognition (ROC)

1 Introduction

In general, AFR can be accomplished through class discrimination which can be realized and implemented through numerous methods and techniques. While some of these techniques are based on deterministic principle, such as the technique of

H. Darwish (✉)
Telecommunication Department, Faculty of Engineering, AlWatania Private University, Hamah, Syria
e-mail: hikmat.darwish@wpu.edu.sy

C.-C. Wang, A. Nallanathan (eds.), *Proceedings of the 5th International Conference on Signal Processing and Information Communications*, Signals and Communication Technology, https://doi.org/10.1007/978-3-031-13181-3_2

discrete Fourier transform (DCT) and Gabor filter [1–3], other AFR techniques can be considered to have statistical basis. They are mainly principal component analysis (PCA) [4–6], linear discernment analysis (LDA) [7, 8], support vector machine (SVM) [9], and artificial neural networks (ANN) [10].

In this paper we are interested in applying two different techniques to AFR, namely, PCA and LDA operating on still face images but subjected to gradient-shadowing being a specific image occlusion. The motivation can be justified by the fact that side-shadowing is the occlusion imposed on face images most often. The paper contains basic theoretical analysis of both techniques, illustrating the advantage gained by LDA over PCA technique in terms of enhanced capability of LDA compared with PCA with regard to class separation. Computer simulation evaluates the performance of each above techniques in terms of recognition rate against increased levels of gradient-shadowing occlusion.

2 Literature Review

AFR which is based on PCA principles was first suggested by M. Kirby and L. Sirovish [6] in 1990, and it was utilized by And M. Turk and A.P. Pentland [7] later on. Aleix M. Martinez and Avinash C. Kak [8] have tested the performance of both PCA and LDA against subspace dimensionality. The simulation results showed a slight improvement of 20% on average in terms of FRR in favor of LDA over PCA. Similar results were obtained by Peng P. et al. [11]. The work of Chandolu P. and Jayesh G [12] illustrate almost 16% on average FRR improvement of LDA performance over PCA when the techniques run against increased number of face samples. Similar results have been obtained by Önsen T. and Adnan A. [13]. Tomesh Verma and Raj Kumar Sahu [14] tested LDA-PCA technique which incorporate PCA feature vectors versus PCA technique alone. Simulation results exhibit once again 16–20% improvement in FRR gained by LDA-PCA over PCA alone. However, the work does not include the effect of face image occlusion on the FRR for both techniques, in addition to the fact that the combined LDA-PCA can prove to be too costly in terms of the required processing speed for real-time application. The work of Zhao W. et al. [15] retains similar result but with different type of face image occlusions.

3 PCA Technique

3.1 Theoretical Background

PCA technique is based on Eigen vector decomposition of data covariance matrix. The information of data, say, that of face image, is represented by a set of most significant Eigen vectors of the covariance matrix. This basic concept of PCA will be explained in the following example. Let X and Y be data vectors such as

$$X = [2.3\ 0.5\ 2.2\ 1.9\ 3.1\ 2.3\ 2\ 1\ 1.5\ 1.1]$$

and

$$Y = [2.4\ 0.7\ 2.9\ 2.2\ 3\ 2.7\ 1.6\ 1.1\ 1.6\ 0.9]$$

The two data vectors are plotted in Fig. 1. The sample means are

$$\overline{X} = \frac{\sum_{i=1}^{10} x_i}{10} = 1.81 \tag{1}$$

and

$$\overline{Y} = \frac{\sum_{i=1}^{10} y_i}{10} = 1.91 \tag{2}$$

The normalized data vectors to the sample means are

$$A = X - \overline{X}$$
$$= [0.69 - 1.31\ 0.39\ 0.09\ 1.29\ 0.49\ 0.19 - 0.81 - 0.31\ 0.71]$$

and

$$B = Y - \overline{Y}$$
$$= [0.49 - 1.21\ 0.99\ 0.29\ 1.09\ 0.79 - 0.31 - 0.81 - 0.31\ 1.01]$$

Then the covariance matrix is

$$C = \begin{bmatrix} 0.61655 & 0.61544 \\ 0.67787 & 0.71655 \end{bmatrix}$$

And the corresponding Eigen vectors are

$$e = \begin{bmatrix} -0.73517 & -0.67787 \\ -0.67787 & 0.73517 \end{bmatrix}$$

The two Eigen vectors represent two PCA components—PCA1 and PCA2—as shown in Fig. 1. In general, we chose Eigen vectors so that data have maximum variance along these vectors to be the PCAs and neglect the rest. Data can be projected on these major PCAs and thus reducing data dimensionality. In our example, we have chosen the first Eigen vector $e^T = [-0.73517\ -\ 0.67787\ ]$ as single PCA and neglected the other.

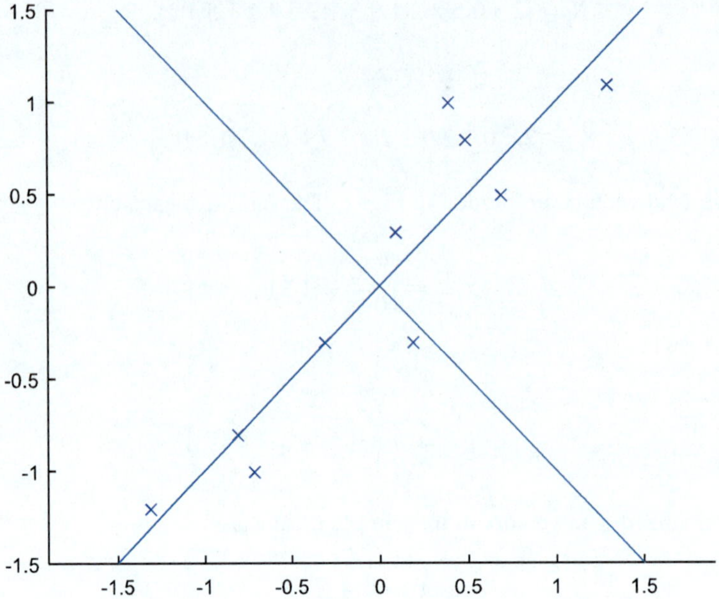

Fig. 1 Ten data points representing *X* and *Y* vectors

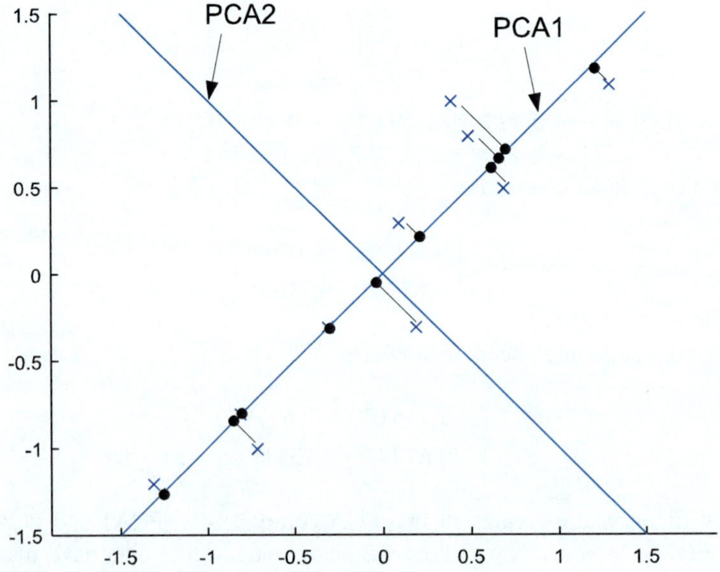

Fig. 2 Projection of data on main PCA vector

When data is projected on this PCA (black dots), the data dimensionality (size) is reduced (compressed) from two dimensions to one dimension as shown in Fig. 2. This is the first function of PCA analysis.

However, and most importantly, these two Eigen vectors represent what is called *feature vectors* of the data, and hence they can be used for recognition in general and for face recognition in particular. Which is considered to be the second benefit of PCA principle.

3.2 *PCA Applied to Face Recognition*

Regardless of the technique being used, AFR is usually accomplished by executing two phases. These are training phase and recognition phase.

3.2.1 Training Phase

Stage 1: The system starts by digitizing the camera snapshot into 2D face image of black/white intensity of size $n \times m$ pixels. Also, in this stage, pixels of the 2D image are concatenated to form 1D image vector X of size $(L \times 1)$ where $L = m \times n$.

Stage 2: Calculating the "average face" which is the averaging of M given faces vectors such that

$$\varphi = \frac{1}{M}\sum\nolimits_{i=1}^{M} X_i \tag{3}$$

Stage 3: Normalizing all available M face images to zero average value by subtracting

$$\psi_i = X_i - \varphi \tag{4}$$

Stage 4: Constructing face covariance matrix C and calculating PCAs through the following:

Construct face image matrix

$$A = \left[\psi_1^T \ \psi_2^T \ldots \ \psi_M^T\right] \tag{5}$$

Calculate covariance matrix

$$C = \frac{1}{M} A A^T \tag{6}$$

where A^T is transpose of matrix A.
Calculate M Eigen vectors of C matrix as

$$e = Av \tag{7}$$

where v is the largest Eigen vectors of matrix:

$$C = A^T A \tag{8}$$

Vectors e of largest corresponding λ are the required PCA for this person image.
Stage 5:
Project face images on PCAs (Eigen vectors) as

$$P_i = e_i^T \psi_i \tag{9}$$

and form person image feature vector

$$\Omega = [P_1 P_2 \ldots . P_M]^T \tag{10}$$

3.2.2 Recognition Phase

Stage 1: Enter query images, repeat stages 1–5, and obtain the feature vector of the query image Ω_x.
Stage 2: Calculate Euclidian distance between query image feature vector and each one of the stored images.

$$d_j = \left\| \Omega_x - \Omega_j \right\| \tag{11}$$

where $J=1,2...,K$, where K is the number of persons in the data base and consequently identify the query image that has the index

$$j = \operatorname{argmin}(d_j) \tag{12}$$

4 LDA Technique

Principal component analysis (PCA) is an effective method mainly for reducing data dimensionality, thus for data compression, and it has been used for face recognition systems effectively. However, PCA technique does not take into account class separation. In other words, there could be two well-separated classes of data having the same spreading direction as shown in Fig. 3, and the PCA technique considers them as single class with one Eigen vector.

Another statistical technique that is called Fisher's linear discernment analysis or LDA for short which was first introduced by Sir Ronald Fisher in 1936, and reformulated later on by S. Balakrishnama and A. Ganapathiraju [16] and Alaa Tharwat et al. [17], provides both Eigen vectors and class separation at the same time; hence, there could be a big chance for the LDA to outperform PCA technique in terms of ROC for a wide verities of ill-illuminated images.

The main principle of LDA technique is to find the Eigen vectors that maximize the separation between classes in the data so that a powerful class recognition system may be obtained.

With LDA technique it may well be possible to obtain both PCA projection and concise class separation at the same time (Fig. 3a, b).

In order to drive mathematical formula for obtaining these projection vectors, let us assume two classes X_1 and X_2. Each class contains M number of the same person's image vectors, of which each is of $N \times 1$ size such as

$X_1 = \left(X_1^1, X_1^2, \ldots, X_1^M\right)$. The mean value for class X_1 is

$$\mu_1 = \frac{1}{M}\sum\nolimits_{i=1}^{M} X_{1,i} \tag{13}$$

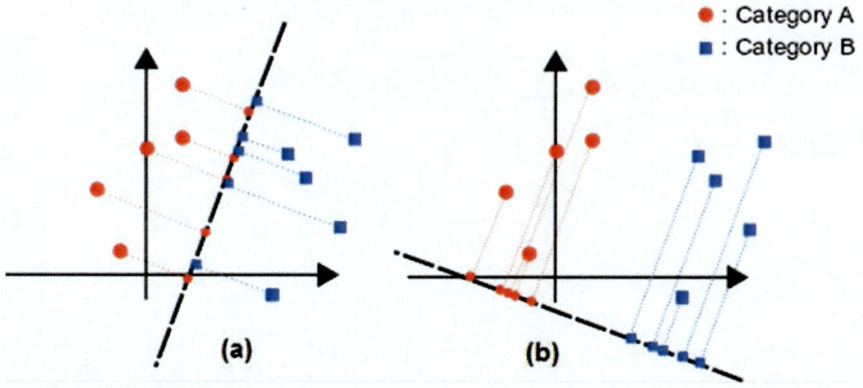

Fig. 3 (**a**) Projection of data on main PCA. (**b**) Projection of data on main LDA showing effective class (category) separation

$$\mu_2 = \frac{1}{M}\sum\nolimits_{i=1}^{M} X_{1,i} \tag{14}$$

And the mean value for class X_2 is

$$\mu_2 = \frac{1}{M}\sum\nolimits_{i=1}^{M} X_{2,i} \tag{15}$$

Note that both μ_1 and μ_2 are vectors on size $N \times 1$.

Let w be the LDA projection vectors that will guarantee the separation of face image classes, and then the mean of the first face image projection on this vector is

$$\widetilde{\mu}_1 = w^T \mu_1 \tag{16}$$

And the mean of second face image projection on this vector is

$$\widetilde{\mu}_2 = w^T \mu_2$$

The projection of class X_1 on vector w is the vector Y_1 such that

$$Y_1 = w^T X_1 \tag{17}$$

And for class X_2, the projection vector is Y_2 such that

$$Y_2 = w^T X_2 \tag{18}$$

The distance between these classes' means is

$$d(w) = |\widetilde{\mu}_1 - \widetilde{\mu}_2| = |w^T \mu_1 - w^T \mu_2| = |w^T(\mu_1 - \mu_2)| \tag{19}$$

The covariance matrix, otherwise known as scatter matrix for projection Y_1 as

$$\widetilde{S}_1 = \sum\nolimits_N (Y_1 - \widetilde{\mu}_1)(Y_1 - \widetilde{\mu}_1)^T \tag{20}$$

And for class Y_2

$$\widetilde{S}_2 = \sum\nolimits_N (Y_2 - \widetilde{\mu}_2)(Y_2 - \widetilde{\mu}_2)^T \tag{21}$$

Rewriting Eq. (20),

$$\widetilde{S}_1 = \sum\nolimits_N w^T (X_1 - \widetilde{\mu}_1)(X_1 - \widetilde{\mu}_1)^T w = w^T \left[\sum\nolimits_N (X_1 - \widetilde{\mu}_1)(X_1 - \widetilde{\mu}_1)^T\right] w \tag{22}$$

Or

$$\widetilde{S}_1 = w^T S_{w1} w \tag{23}$$

where

$$S_{w1} = \sum\nolimits_N (X_1 - \widetilde{\mu}_1)(X_1 - \widetilde{\mu}_1)^T \tag{24}$$

is called within-class scatter matrix.

Then

$$\widetilde{S}_1 + \widetilde{S}_2 = w^T S_w w \tag{25}$$

where

$$S_w = S_{w1} + S_{w2} \tag{26}$$

is again called within-class scatter matrix.

Now in order to obtain the required LDA projection vector w which will separate the two classes, we seek the vector w that maximizes the ratio of mean distance to within-class matrices:

$$\rho(w) = \frac{|\widetilde{\mu}_1 - \widetilde{\mu}_2|^2}{\widetilde{S}_1 + \widetilde{S}_2} \tag{27}$$

On the other hand, rewriting

$$|\widetilde{\mu}_1 - \widetilde{\mu}_2|^2 = (w^T\mu_1 - w^T\mu_2)^2 = w^T(\mu_1 - \mu_2)(\mu_1 - \mu_2)^T w \tag{28}$$

$$|\widetilde{\mu}_1 - \widetilde{\mu}_2|^2 = w^T S_B w \tag{29}$$

where

$$S_{Bi} = (\mu_i - \mu_m)(\mu_i - \mu_m)^T \tag{30}$$

is called the between-class scatter matrix.

In general and for N face images, the between-class scatter matrix is given by

$$S_B = \sum\nolimits_{i=1}^{N} (\mu_i - \mu_m)(\mu_i - \mu_m)^T \tag{31}$$

Then the ratio to be maximized in order to get class separation becomes

$$\rho(w) = \frac{w^T S_B w}{w^T S_w w} \tag{32}$$

The LDA-required projection vector is obtain by

$$w = \text{argmax}\left(\frac{w^T S_B w}{w^T S_w w}\right) \tag{33}$$

And this can be carried out by doing

$$\frac{d\rho(w)}{w} = 0 \tag{34}$$

And this leads to

$$S_w^{-1} S_B w = \rho w \tag{35}$$

The interpretation of Eq. (35) is that the required LDA projection w vector is the Eigen vector of matrix

$$A = S_w^{-1} S_B \tag{36}$$

which can be obtained in similar manner to PCA technique.

Example:

Let two images be represented by the two vectors:

$$X_1 = \{(4, 1), (2, 4), (2, 3), (3, 6), (4, 4)\}$$
$$X_2 = \{(9, 10), (6, 8), (9, 5), (8, 7), (10, 8)\}$$

Then we get

$$S_1 = \begin{bmatrix} 0.8 & -0.4 \\ -4.0 & 2.6 \end{bmatrix}, S_2 = \begin{bmatrix} 1.84 & -0.04 \\ -0.04 & 2.64. \end{bmatrix}$$

and

$$\mu_1 = [3.00\ 3.60],$$
$$\mu_2 = [8.40\ 7.60]$$

Substituting in Eqs. (35) and (36), we get

$$S_w = \begin{bmatrix} 2.64 & -0.44 \\ -0.44 & 5.28 \end{bmatrix},$$

$$S_B = \begin{bmatrix} 29.16 & 21.60 \\ 21.60 & 16.00 \end{bmatrix}$$

$$S_w^{-1} S_B = \begin{bmatrix} 11.98 & 8.81 \\ 5.08 & 3.76 \end{bmatrix}$$

Solving Eq. (35) we get

$$S_w^{-1} S_B w = \lambda w \longrightarrow \left| S_W^{-1} S_B - \lambda I \right| =$$

$$\begin{vmatrix} 11.98 - \lambda & 8.81 \\ 5.08 & 3.76 - \lambda \end{vmatrix} = 0 \longrightarrow \lambda = 15.65$$

Then

$$\begin{bmatrix} 11.98 & 8.81 \\ 5.08 & 3.76 \end{bmatrix} \begin{bmatrix} w_1 \\ w_2 \end{bmatrix} = 15.65 \begin{bmatrix} w_1 \\ w_2 \end{bmatrix}$$

Solving for w_1, w_2

$$\begin{bmatrix} w_1 \\ w_2 \end{bmatrix} = \begin{bmatrix} 0.91 \\ 0.39 \end{bmatrix}$$

Figure 4 illustrates the two image classes and the LDA projecting vector on which the projections of two images can be easily separated.

5 Computer Simulation

A computer simulation for evaluating the performance of both PCA and LDA techniques using the same images have been implemented through writing MatLab program which was run on laptop computer core i5 machine.

The simulation consists of processing face images of 100 different person samples. Each face image is represented by 200 × 200 pixels of monochromatic gray intensity, a sample of which is shown in Fig. 5.

These face images have been borrowed with courtesy from free image library of the University of Sheffield [18].

The simulation consists of both training and recognition phases for each of PAC and LDA techniques. In training phase, the data base which contains feature vectors are obtained using ideal images regarding the lighting conditions. Only in

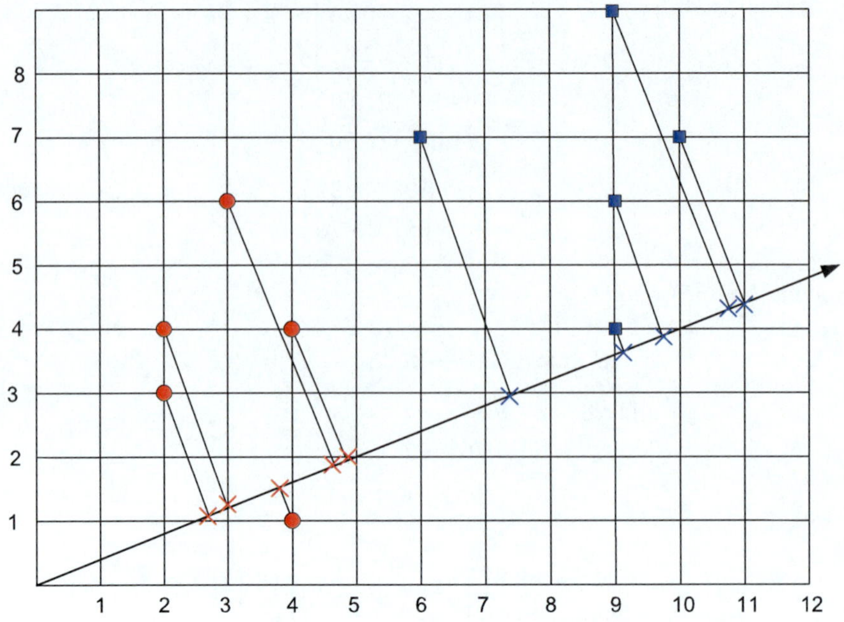

Fig. 4 Projection of two simple images on the main LDA vector showing clear class separation

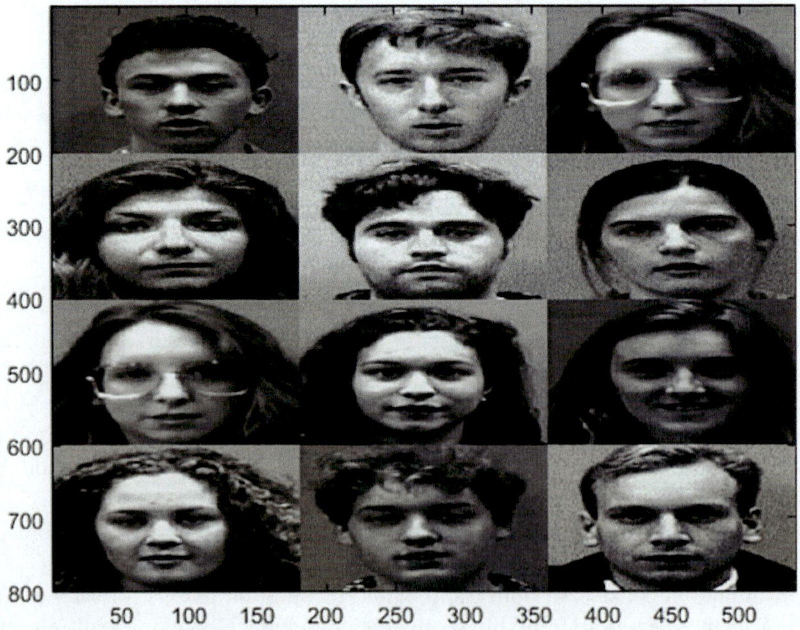

Fig. 5 Face samples taken from the University of Sheffield

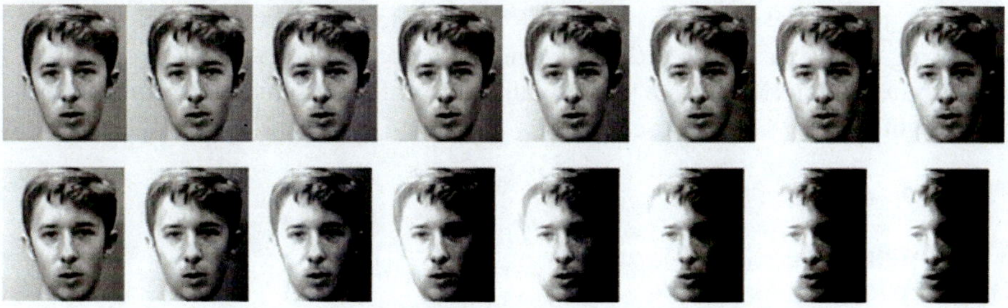

Fig. 6 Side-shadowing occlusion of deferent qualitative grades: (**a**) multi = 0.3 and (**b**) multi = 0.6

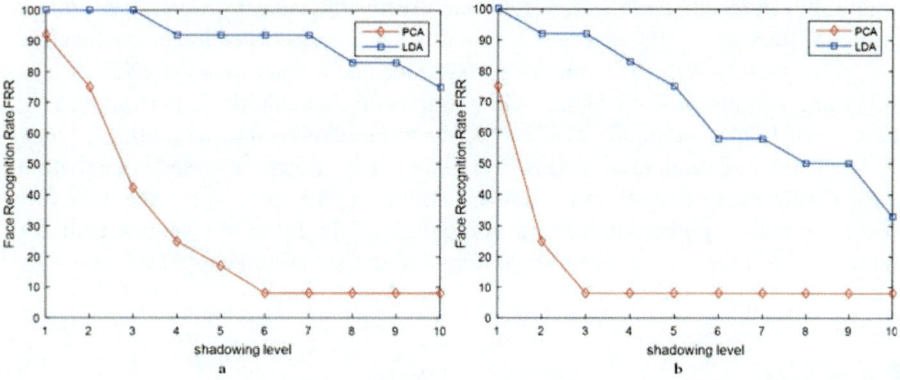

Fig. 7 Face recognition rate of PCA versus LDA. (**a**) multi = 0.3 and (**b**) multi = 0.6

recognition phase that occlusion are introduced over images for testing. Since this work is mainly concerned with relative comparison of PCA and LDA, then qualitative image occlusion might be a sensible choice. Having said that, within the recognition phase, the nonideal (*shadowing level*), which is acting as image occlusion, was introduced by adding a black hue of linear gradient to each image. Figure 6a, b illustrates the increased black hue on two separate experiments with multiplying factor multi = 0.3 and multi = 0.6 respectively.

Each of PCA and LDA algorithms is performed on the same corrupted images set after adding "salt-and-pepper" noise so that we get random *ensample* which is necessary to calculate the FFR which is in turn is a probability. The performance of both PCA and LDA was evaluated in terms of FRR against increased hue intensity (multiplying factor) which is denoted by *shadowing level* as it is depicted in Fig. 7a, b. Figure 7a depicts the simulation outcome for average black hue of multi = 0.3, while Fig. 7b depicts simulation outcomes for multi = 0.6. By studying these figures, we notice the following: first, for ideal illumination, i.e., shadowing =0.

We may notice that the improvement of FRR gained by LDA over PCA is ranging between 10% and 25%, which is in complete agreement with the works of [19–21]. Secondly, as the severity of occlusion increase, the performance in terms of FRR of PCA deteriorates rapidly, while that of LDA deteriorates gracefully. Even we get fairly constant and good FFR of 80% for mild shadowing (Fig. 7b). Thirdly, still LDA technique maintains higher FRR compared with PCA for all levels of shadowing occlusions. From FRR curves in Fig. 6a, b, we notice that LDA technique outperforms PCA techniques in terms of higher FRR for the same degree of nonideal image lighting.

6 Conclusion

In principle, theoretical formula which may enable us to quantify the difference in performance between PCA and LDA in terms of FRR is needed. However, this task can be proved to be highly difficult. Monte Carlo simulation as it was used in this work may provide a valid alternative. And as such, we succeeded in getting tangible indication that LDA can outperform PCA technique in terms of FRR versus ill-lighting condition. Both figures show a graceful degradation in performance in the case of LDA techniques with respect to worsening lighting conditions, while PCA fails fast and dramatically. However, the price is a much increased computation burden in the case of LDA, and this might not be a prohibiting factor given the fact that computation power of modern computers might fulfill the requirement for implementing LDA-based face recognition system for real-time applications.

References

1. C. Liu, H. Welchsler, Gabor feature classifier for face recognition. Proc. ICCV **2**(5), 270–275 (2001)
2. B. Kepenekci, F.B. Tek, G.B. Akar, Occluded face recognition by using Gabor features, in *3rd COST 276, Workshop on Information and Knowledge Management for Integrated Media Communication, Budapest*, (2002)
3. T.M. Abhishree, J. Latha, K. Manikantan, S. Ramachandran, Face recognition using Gabor filter based feature extraction with anisotropic diffusion as a pre-processing technique. Procedia Comp. Sci. **45**(C), 312–321 (2015). https://doi.org/10.1016/j.procs.2015.03.149
4. K. Pearson, On lines and planes of closest fit to systems of points in space. Philos. Mag. **2**, 559–572. University College London (1901)
5. H. Hotelling, Analysis of a complex of statistical variables into principal components. J. Educ. Psychol. **24**, 417–441 and 498–520 (1933)
6. M. Kirby, L. Sirovish, Application of the Karhunen-Loeve procedure for the characterization of human faces. IEEE Trans. Pattern Anal. Mach. Intell. **12**, 103–108 (1990)
7. M. Turk, A.P. Pentland, Eigen faces for recognition. J. Cogn. Neurosci., 71–86 (1991)
8. A.M. Martinez, A.C. Kak, PCA versus LDA. IEEE Trans. Pattern Anal. Mach. Intell. **23**(2), 228–233 (2001)
9. G. Guo, S.Z. Li, K. Chan, Face recognition by support vector machines, in *Proceedings Fourth IEEE International Conference on Automatic Face and Gesture Recognition (Cat. No. PR00580), 28–30 March 2000*,
10. S.A. Nazeer, N. Omar, Marzuki Khalid Face Recognition System using Artificial Neural Networks Approach, 2007 International Conference on Signal Processing, Communications and Networking

11. P. Peng, I. Portugal, P. Alencar, D. Cown, *A Face Recognition Software Framework Based on Principal Component Analysis* (University of Waterloo, Waterloo, 2021)
12. C. Prasanthi, J. Gangrade, Face recognition for different facial expressions based on PCA, LD analysis. Int. J. Dev. Res. **07**(07), 14109–14112 (2017)
13. Ö. Toygar, A. Acan, Face recognition using PCA, LDA and ICA approaches on colored images. J. Electr. Electron. Eng. **3**(1), 735–743 (2003)
14. T. Vermaand, R.K. Sahu, PCA-LDA based face recognition system & results comparison by various classification techniques, in *Proceeding of International conference on Green High Performance Computing, India 2013*,
15. W. Zhao, R. Chellappa, A. Rosenfeld, P.J. Phillips, Face recognition: A literature survey, in *Technical Report CAR-TR-948, CS-TR-4167, N00014-95-1-0521*, (October 2000)
16. S. Balakrishnama and A. Ganapathiraju. Linear discriminant analysis-a brief tutorial. Institute for Signal and information Processing, 1998
17. A. Tharwat, T. Gaber, A. Ibrahim, A.E. Hassanien, *Linear Discriminant Analysis: A Detailed Tutorial* (Ai Communications, 2017)
18. https://www.sheffield.ac.uk/eee/research/iel/research/face
19. F. Mahmud, M.T. Khatun, S.T. Zuhori, S. Afroge, M. Aktar, B. Pal, Face recognition using principle component analysis and linear discernment analysis, in *International Conference on Electrical Engineering and Information Communication Technology (ICEEICT)*, (2015)
20. J. Yang, H. Yu, W. Kunz, "Linear Discernment Analysis" school of computer science interactive systems laboratories Carnegie Mellon University Pittsburgh, pa 15213
21. P.N. Belhumeur, J.P. Hespanha, D.J. Kriegman, Eigenfaces vs fisherfaces: Recognition using class specific linear projection. IEEE Trans. Pattern Anal. Mach. Intell. **19**(7), 711–720 (1997)

The Effect of the Audio Signals on the Brain for EEG-Based Attentiveness Recognition

Jiaming Cai, Pinzhen Chen, Yunyang Yuan, and Zecheng Li

Abstract Nowadays, people are living in a world filled with all kinds of voice and noise. However, when trying to focus on work, the external noise and sounds will interrupt and prevent us from concentrating, thus, this paper investigates the neural activities related to the working attentiveness by using EEG (electroencephalography). The purpose is to use the collected data to build an EEG-based BMI (brain-machine interface) to detect which kind of noise is distracting the person and then use this information to shield the correlated noise or create another kind of audio to cancel the noise. Our work indicates that the level of people concentrating on their work will decrease rapidly when the outside noise reaches a certain level. And with the help of technology, it is able to detect the changes of people's attentiveness and then choose whether to block the outside noise to increase the level of attention.

Keywords Brain-machine interface (BMI) · Electroencephalography (EEG) · Alpha wave · Beta wave · Attention

J. Cai
College of Engineering, University of Kentucky, Lexington, KY, USA

P. Chen (✉)
School of Electronic Science & Engineering, Southeast University, Nanjing, China
e-mail: 213190350@seu.edu.cn

Y. Yuan
School of Communication and Information Engineering, Nanjing University of Posts and Telecommunications, Nanjing, China

Z. Li
School of Electrical and Computer Engineering, Ohio State University, Columbus, OH, USA

C.-C. Wang, A. Nallanathan (eds.), *Proceedings of the 5th International Conference on Signal Processing and Information Communications*, Signals and Communication Technology, https://doi.org/10.1007/978-3-031-13181-3_3

1 Introduction

1.1 Brain-Machine Interface and EEG

Brain-machine interface (BMI) is a relatively burgeoning technology that links the human brain with outside equipment to enable users to communicate or control the external equipment by using stimulated signals [1]. Our brain-machine interface analyzes the signals collected by electroencephalography (EEG), which records the neural activity and transfers them to visible figures to determine what command should be executed next to complete the purpose.

1.2 Attention

As we all know, the external sound and noise will affect human beings, including their emotions, activities, and attention. Humans' attention can be affected by different sound frequencies, which can be analyzed by EEG. There are research about the effects of classical music on people's attention through EEG technology, and there is a well-known phenomenon called the "Mozart effect" [2].

Our goal is to show participants' attentiveness by analyzing EEG signals and to find the appropriate periods of frequency of audio signals to be blocked. Thus, people can enhance their level of attention without the external noise and become more efficient on their tasks.

Most of neural activities emitted from our brain can be transformed into the form of EEG signals. After amplifying and filtering the raw EEG signal into different frequencies, one can access humans' attentiveness. There are several basic types of brain waves which are called alpha, beta, delta, and theta waves. Alpha waves range from 8 Hz to 14 Hz, and it only occurs when individuals are highly calm. People who are meditating, reading, or awake might generate strong alpha waves. Theta waves range from 4 Hz to 7 Hz and only exist while people are sleeping. Delta waves have the lowest frequency, which is under 3.5 Hz, and only occurs during deep sleep. Beta waves range from 12 Hz to 30 Hz, and it's the reflection of people's attention [3]. Moreover, beta waves will be in higher abundance when a participant is focused on some types of tasks or thinking. Thus, it is the main type of waves this study will focus on.

2 Methodology

This experiment sets a blank control group and two experimental groups and recruited 20 healthy subjects aged 18–25, including 10 boys and 10 girls. They are college students, with normal hearing and normal or corrected vision. All signed an informed consent before the experiment and received after the experiment a certain reward.

2.1 Specification of Experiment

The experiment was carried out in a shielded room. The subjects were 50 cm away from the computer, kept in a comfortable sitting position, and looked at the center of the screen. Ten subjects (5 boys and 5 girls) were randomly selected from the 20 subjects for a blank control experiment. Each subject was required to read two English test documents, each of which was between 200 and 300 words. The reading time was 5–6 minutes, with a 1–2-minute break between the two readings. Meanwhile, their brain waves were recorded.

In the two experimental groups, the remaining ten subjects read the same two English articles as the blank control group in the same test environment. Each group read the article for 5–6 minutes, with a rest of 1–2 minutes between the two groups. In experimental group 1, no audio interference was applied during the first minute then from the second minute an audio interference was applied from minute 2 until the end of reading. In experimental group 2, no audio interference source was applied in minute 1, and another given audio interference source 2 was applied in minute 2 until the end of reading. The brain waves of the two groups were recorded. Before the formal experiment, each subject was given a practice test of ten times. During the experiment, subjects were asked to keep their heads still and blink as little as possible.

Below are the specifications of the audio.

Audio interference source 1: No audio interference source was applied in 1 minute.

Audio Interference Source 2: Based on the audio interference source 1, from 0 to 10s, a combination of ambient noise and fixed-frequency audio, including random noise sources, rushing cars, airplanes, people's noise, and other sources of ambient noise. Fixed audio frequencies that were used are 50–200 Hz, 400–600 Hz, 1000–1300 Hz, and 3000–3200 Hz. When it comes to the specific period of time, human is used to seeing difference in magnitude of brain waves, remove mixed audio frequencies of 50 Hz from 10s to 20s, solid audio frequencies of 400 Hz from 20s to 30s, solid audio frequencies of 1500 Hz from 30s to 40s, and solid audio frequencies of 3000 Hz from 40s to 50s.

One thing that needs to be noticed is that after reading of the blank control group, mean-variance analysis was performed on the concentration values of the subjects after two readings. If the variance was too large, the blank control group was added, and another group of English articles was read until two articles with similar concentration for the same subject were screened out.

2.2 Equipment of the Experiment

In our experiment, we used the NeuroSky MindSet headset as a data collection device as shown in Fig. 1. Traditionally, electromagnetic waves are collected according to the international 10–20 electrode placement system (10–20 system), which involves 37 electrodes on the skull. However, human beings' level of attentiveness are only controlled by the cerebral cortex in the forehead at Fp1; applying the tradition 37-electrode headset to testers is inconvenient and impractical to the research. Though the device carried one dry sensor at Fp1 on the skull as shown in Fig. 2, it can fulfill our experiment's setup. And the reference sensor is placed on the testees' left ear. This device also attached with a research tool box including NeuroView software and Neuroskylab MATLAB module, which can make it easier to connect, graph, view, and record the signals in real time. The NeuroSky MindSet Research Tools (MRT) is cost effective, portable, and user-friendly. This device uses a ThinkGear AM chip module technology to identify and digitize weak EEG signals and filter the electrical signal from the surrounding, but it will also detect the EMG signal on the forehead. So the muscle movement of the testees should be controlled to simplify the data analyse later. It will collect, filter, process, and analyze EEG signals and deliver processed and digital EEG signals in the alpha, beta, delta, gamma, and theta bands at 512 Hz sample rates. The headset is connected to the computer through the wireless Bluetooth connection, so we ensured all subjects were sitting at the same distance from the computer during the experiment to avoid the errors of the signal amplitude. The detected signal was then sent to the NeuroView platform or the MATLAB Neuroskylab module, as the example in Fig. 3.

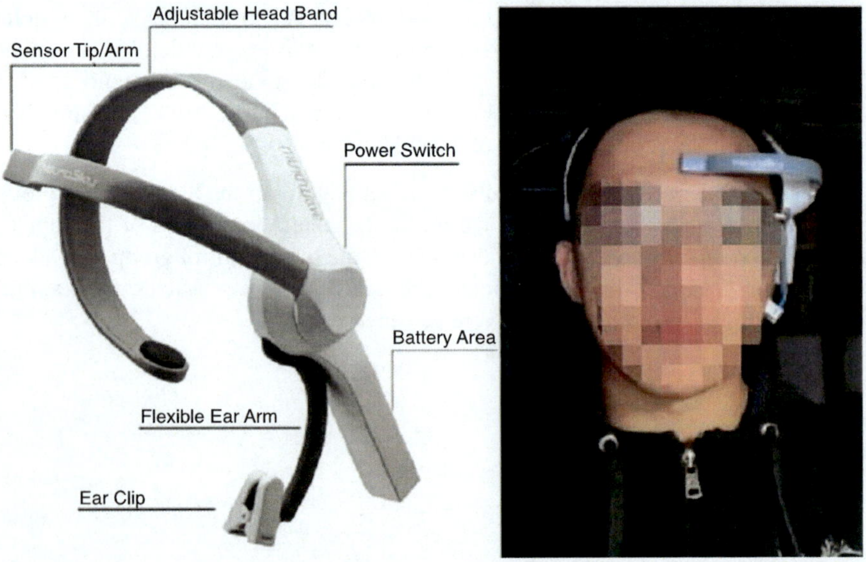

Fig. 1 The Neurosky MW1000 device design and application method

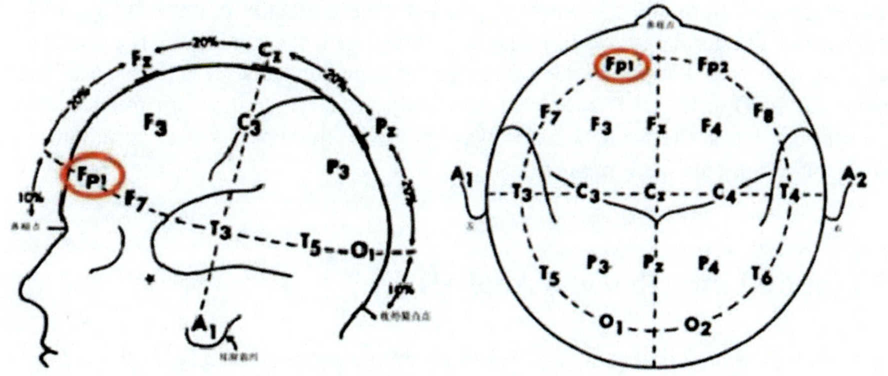

Fig. 2 International 10–20 electrode placement system

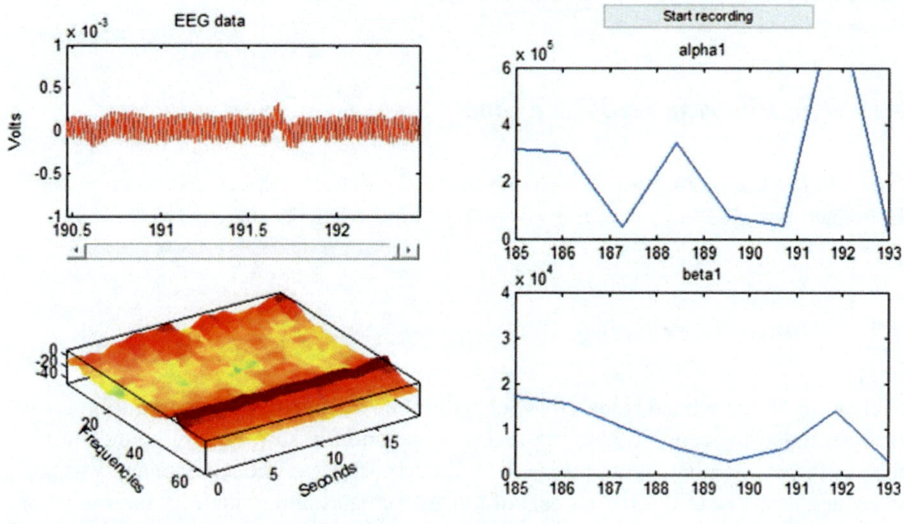

Fig. 3 The collected data on MATLAB Neuroskylab module

3 Data Acquisition and Preprocessing

The NeuroScanSynAmps2 system was used to collect EEG signals from the subjects. The system amplified and bandpass filtered (0-100 Hz) the signal prior to sampling it at 512 Hz. The impedance between scalp and electrode was less than 5KOhm. The behavioral data (whether the subjects responded to the key and the time from the beginning of the stimulus to the response) were recorded using E-Prime 3.0. Before ERP analysis, the collected EEG data needed to be preprocessed by EEGlab as follows: the reference potential was converted to the average value of the left and right mastoid processes, and the original signals were bandpass filtered at 0.1 HZ–30 Hz. The filtered signals were divided into three groups according to the time of stimulus presentation, and the wrong segments were manually eliminated by combining with behavioral data. At the same time, the segment containing large artifacts was also removed. Finally, independent component analysis (ICA) was used to remove the EEG artifacts. However, in the study of classification performance, the 8th-order bandpass filter of 0.1–20 Hz was applied to the data 1 s after stimulation, and all data segments were retained.

4 Feature Extraction and Analysis

For data analysis, the raw data, which is at a sampling rate of 512 Hz with a 15-bit quantization level [4], needs to be preprocessed before analysis; there are two ways to process the raw data: by using the filter provided by EEGlab to sift a raw data or delete the spikes caused by other movements such as blink manually.

4.1 The Filtering in the EEGlab

We used the filtering option in EEGlab to get the data in a specific frequency band associated with the α, β, δ, θ,and γ signals, with the help of the low-pass filter.

4.2 Manually Filtering

According to the related research [4] using mobile sensors to detect attention, it was recommended to recorded also the facial image and the surrounding sounds at the time of the experiment. This is done to facilitate post-experimental data analysis and to accurately record the EEG signals of the test subjects and to identify their level of attentiveness. After comparing the data graphs with facial expressions, it was easy to find out the spikes or fluctuations caused by other factors.

4.3 Data Analysis

We used FFT (fast Fourier transform),

$$X_k = \sum_{n=0}^{n-1} x_n e^{-i2\pi kn/N} \quad k = 0, \ldots, N-1 \tag{1}$$

to convert the original signal into individual spectral components and have the graph which contains frequency information. Therefore, it is easy to tell if there is a dominant frequency that exists when applying a special frequency on the subjects. Moreover, it is necessary to calculate the average amount of the amplitude for it as a reflection of attention as an example. Finally, the attention should be defined by many factors including the average of the amplitude for the EEG signals, the dominant amplitude, and the ratio between α waves and β waves.

5 Discussion

5.1 Mixed Signals

In this study, all the EEG signals are detected from only one electrode—Fp1—on the forehead, which makes the subjects feel more comfortable with the experiment. However, using only a single electrode sacrifices the accuracy because every single facial expression would have a magnificent effect on the EEG signal. The signal extracted by the machine is a mixture of EMG and EEG signals. As a result, any blink of the eye will affect the waveform of the signal, as shown in the Fig. 4, which clearly demonstrates the dramatic changes in waveform amplitudes when subjects blink. To solve this problem, filtering out EMG signals with the help of EEGLAB is indispensable.

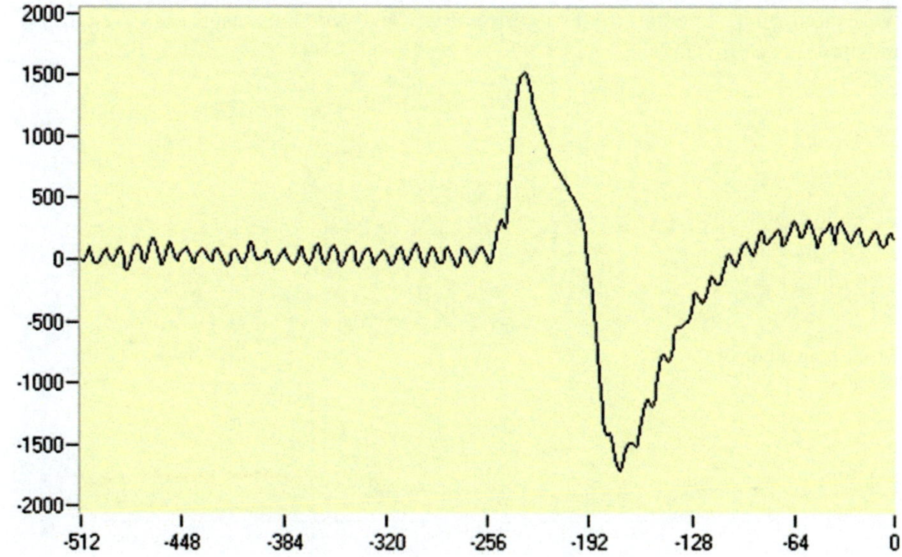

Fig. 4 Influence of eyeblink on signal

Thus, during data analysis, the first step is to filter out all the noise beyond 50 Hz [5]. Using EEGlab, the absolute value of the audio waveforms with a power-law compression with exponential of 0.6 was taken to obtain the audio envelopes, followed 8 Hz low-pass filtering [6]. Envelopes were extracted from the clean audio signals for each speakers [7]. Because of the low accuracy of the experiment equipment, it was necessary to analyze more data and then get the conclusion to improve accuracy and credibility.

By taking the average, checking the dominant frequency for each group and comparing the data with other groups, it was easy to find out the wrong data.

5.2 *Distance*

It's confusing that the amplitude of the signal could be easily affected by the distance between respondents and the computer which is used to process signals. Making sure that the distance between the respondents and computer could be at a specific value before testing is a useful way to avoid these signal variabilities.

5.3 *Insensitivity of the Equipment*

The graphs shown in Figs. 5 and 6 is the magnitude of attention measured by a software called NeuroView. Due to the lack of sensitivity of the equipment, the value of the attention could occasionally be a very low value and affect the testing result. These abnormal statistics made it difficult to analyze data, as a set of example statistics shown in Table 1.

Fig. 5 Magnitude of attention

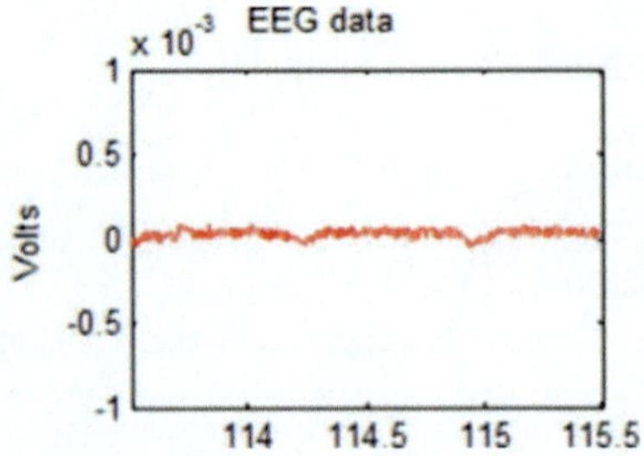

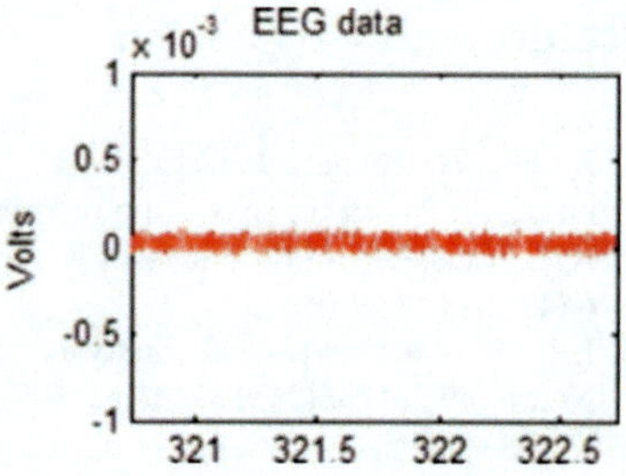

Fig. 6 Magnitude of attention

Table 1 Abnormal statistics during experiment

14:08:51	17
14:08:52	13
14:08:53	1
14:08:54	1
14:08:55	1
14:08:56	14

6 Conclusion

To summarize, when all periods of frequency are shown, at the first 10 seconds, the magnitude of beta wave is around 1 (unit of 10^4), and it doesn't change much along as time goes by, which means that people's attention is low. However, when it comes to the next 20 seconds, as the 50–600 Hz period of frequency of the audio is blocked, the magnitude of beta waves increases above 4 (unit of 10^4). After 30s, the magnitude of beta wave drops rapidly, which shows that user's attention lowers. Thus, people's attention are much higher when low-frequency periods of the audio are blocked.

When looking at the whole picture of this research, the benefits of figuring out how the brain reacts to different kinds of external stimuli at the same time can be detected. The first half of BMI structure is collecting and processing the data, which helped us determined what stimulus triggers what kind of reaction in our brain wave. After understanding the principle of such neural activities, the stimulation of our brain waves can be improved. Currently, this study is only aiming at doing the first half of the brain-machine interface due to the restriction of equipment and areas. Thus, the further study are going to complete the study. After figuring out how is the best way that make people have higher attention, this study can start to construct the second half of the BMI project. The finished product will be like a headphone or earphones but with the function that can block the specific frequency of the noise, helping users to focus on their work. When the signal collector detects that user's beta waves are dropping, it will automatically block the noise that affects the user.

Acknowledgments The authors contributed equally to this work and should be considered co-first authors. The authors would like to thank Professor Jan Van der Spiegel and Ruoyan Zhang for useful suggestions and discussions.

References

1. B. He, B. Baxter, B.J. Edelman, C.C. Cline, W.Y. Wenjing, Noninvasive brain-computer interfaces based on sensorimotor rhythms. Proc. IEEE **103**(6), 907–925 (2015)
2. F.H. Rauscher, G.L. Shaw, N.K. Catherine, Music and spatial task performance. Nature365 **6447**, 611 (1993)
3. J.P. Banquet, Spectral analysis of the EEG in meditation. Electroencephalogr. Clin. Neurophysiol. **35**(2), 143–151 (1973)
4. N.-H. Liu, C.-Y. Chiang, H.-C. Chu, Recognizing the degree of human attention using EEG signals from mobile sensors. **13**(8), 10273–10286 (2013). https://doi.org/10.3390/s130810273. Received: 5 June 2013/Revised: 25 July 2013/Accepted: 6 August 2013/Published: 9 August 2013
5. A. de Cheveigne´ and D. Arzounian, "Robust detrending, rereferencing, outlier detection, and inpainting for multichannel data," NeuroImage, vol. 172, pp. 903–912, 2018
6. W. Biesmans, N. Das, T. Francart, A. Bertrand, Auditory-inspired speech envelope extraction methods for improved EEG-based auditory attention detection in a cocktail party scenario. IEEE Trans. Neural Syst. Rehabil. Eng. **25**(5), 402–412 (2016)
7. R. Zink, S. Proesmans, A. Bertrand, S. Van Huffel, M. De Vos, Online detection of auditory attention with mobile EEG: closing the loop with neurofeedback. https://doi.org/10.1101/218727

EMG-Based View Controller Using VR Applications

Jingcheng Zhao, Anjun Zhang, Tianao Zeng, Hao Cai, and Yanlong Wang

Abstract Electromyography (EMG) signals can detect the movement of muscles, which can be a method to be applied in video games and the rehabilitation field. In this paper, a system to control view in three-dimension (3D) and virtual reality (VR) is designed. Three muscles were selected palpation to obtain EMG signals hardware. Next, the classification algorithms were used on getting features under five different head-moving directions. Comparing and analyzing electrodes at different muscle positions, signal processing, and classification algorithms, and the accuracy of the initial data results are around 96.5%. For testing, in Unity3D in the real-time model is created to prove data accuracy; through the model, a time-limited game to control the rotation of the head is designed, and all experimental results will be presented intuitively in the form of scores.

Keywords EMG · Classification · Visual reality · Game control · Rehabilitation

J. Zhao (✉)
College of Electronic and Information Engineering, Nanjing University of Aeronautics and Astronautics, Nanjing, People's Republic of China

A. Zhang
International Baccalaureate Division, Ningbo Huamao International School, Ningbo, People's Republic of China

T. Zeng
College of Engineering, University of California, Santa Barbara, CA, USA

H. Cai
College of Engineering, University of Edinburgh, Old College, Edinburgh, Scotland

Y. Wang
College of Electrical and Electronics Engineering, Xiamen University Malaysia, Sepang, Malaysia

C.-C. Wang, A. Nallanathan (eds.), *Proceedings of the 5th International Conference on Signal Processing and Information Communications*, Signals and Communication Technology, https://doi.org/10.1007/978-3-031-13181-3_4

1 Introduction

1.1 Motivation

Video games have been widely popularized in today's public life. As an essential entertainment and social means, its form constantly changes with hardware and software technology development. One form is that it used to be limited to the performance of 2D pictures, but now it can use computers for the 3D real-time calculation and rendering, which is more and more realistic. The second is that the control means to become more and more abundant, such as crucial rocker gyroscope and gravity induction. The third is the evolution of various forms. There have been games launched for fitness (Nintendo® Fitness Boxing™, Ring Fit Adventure™, etc.) that have become a carrier, providing exercising purpose. The continuous renewal of game equipment has derived the function of providing entertainment.

Rehabilitation training can effectively train body balance, accelerate postoperative recovery, and effectively recover neurological injury. With technological progress, rehabilitation training has also derived various ways and means such as physical equipment assistance and electronic equipment support. All these make rehabilitation training more and more intelligent and scientific. As a new method of rehabilitation training, the game has been applied to different rehabilitation training. The most significant advantage of the application of games in rehabilitation training lies in its interest, which can psychologically promote patients' enthusiasm and obtain better results.

Meanwhile, in rehabilitation training or paralyzed people, establishing contact between vision and muscles through games will better stimulate the muscle groups that need to move to achieve the effect of rehabilitation training. In addition, video games facilitate the inclusion of scoring mechanisms so that medical staff and patients can better evaluate the rehabilitation process.

The EMG signal can be effectively sampled and identified as a signal reflecting muscle contraction intensity applied to clinical and industrial fields [1]. Because of its noninvasive nature, it can easily measure the EMG signal of specific muscle groups and realize man-machine interaction. At the same time, it can be used as a new control means to develop games with this as the core.

Based on the above reasons, the combination of EMG signal and game has excellent prospects in rehabilitation training. Games can also get new inspiration through EMG to help developers create new game forms.

Therefore, the sampling of EMG signals of muscles swinging up, down, left, and right in the neck is used to control VR games to become a new form of games or help rehabilitation training. Based on the EMG signal of the neck muscle, this paper applies it to the control of VR scene through machine learning and designs a set of systems.

1.2 Background

At present, the research on EMG has been relatively perfect [2, 3]. In terms of hardware, the signal received by the sampling electrode of the ad chip is generally uploaded to the microprocessor for processing. The obtained data is usually processed in the way of classification. After filtering, normalization, and other operations, the features are extracted in the time domain and frequency domain [4, 5]. Classify the extracted eigenvalues, and get the model to judge the corresponding action. There are also application examples of attitude estimation through the direct application of features through algorithms and mathematical models [6].

The application of classification has good consistency with the recognition of EMG signals. According to the characteristics, the operation is carried out through the classification algorithm [7]. With the support of the above tools, researchers have developed applications based on EMG signals. In [8], a kind of operation gesture recognition similar to pad based on hand motion is proposed. And [9] provides a design of a remote-control cars using gesture control. EMG can better support the development of such applications. The following researchers combined a 3D scenario with EMG signals as a human-computer interaction system [10, 11]. However, most of this research concentrates on hand gestures and arm movements. The investigations on head movement are also developed [12]. chooses frontalis and muscles on the neck to acquire EMG signals. In [13], the author combines inertial measurement unit (IMU) signals with EMG to classify. The two types of research used subsidy methods to raise accuracy and predict gestures with orientation angles. In the paper [10], capture the EMG signal of the hand skeletal muscle and feedback the real-time 3D image to the user through the game development engine with unity. At the same time, the muscle signal collection system uses a differential amplifier to amplify the weak muscle signal to improve the accuracy [14]. uses large numbers of electrodes distributed on the neck to acquire relatively complete movement data. However, the number of electrodes will affect the practicability of the system, which is essential in daily applications.

Based on the research referred above, we developed a system to control view in 3D scenario with neck EMG signals. Compared with existing research, we chose an electric scheme to balance costing, performance, and practicability. Firstly, we use neck muscles only to get three channels of EMG signals. On the hardware, we choose single-chip microcomputer combined with A/D modules to acquire EMG signals. Secondly, we build a complete system including hardware, data collecting, processing, and classification system. Thirdly, we develop a 3D game to assist in demonstrating. It can also assess how the system performs when facing first-time users. Considering that the contraction strength of the neck muscles of different users may differ during the test, we selected several volunteers to test our system. After the continuous screening, high-precision muscle signals will be imported into the game model. The game model is built by Unity, which is intuitive and straightforward; patients can obtain rehabilitation training directly through the game. At the same time, high-precision muscle signals will make the model very sensitive to the user's head movement and almost wholly synchronize the user's actions, making the game process very smooth.

1.3 Report Structure

The remainder of this paper is subdivided as follows: Section 2 describes the materials used, the design of the game scene, and evaluation methods; Section 3 shows the verification of control performance and the results obtained; Section 4 discusses the results; and Sect. 5 is the conclusion drawn from the paper.

2 System Design

The system flow is shown in Fig. 1. The system includes EMG signal collection hardware, data process, and VR client.

2.1 EMG Signal Collection

A six-channel sensor was used to collect EMG signals.

The muscle tissue was connected by three conductive gel electrodes (Shanghai LITU Medical Appliances Co., Ltd) per channel. Two of the electrodes are responsible for EMG signals, and the other one connects to a no-activity muscle which represents reference potential. The three-electrode pairs are placed on selected muscles. The ground electrodes are placed as in Fig. 2.

The sensor could be amplified by Arduino UNO's analogue-to-digital (A/D) pins and transport six-channel data with serial through USB connector at a sampling rate of 1000 Hz.

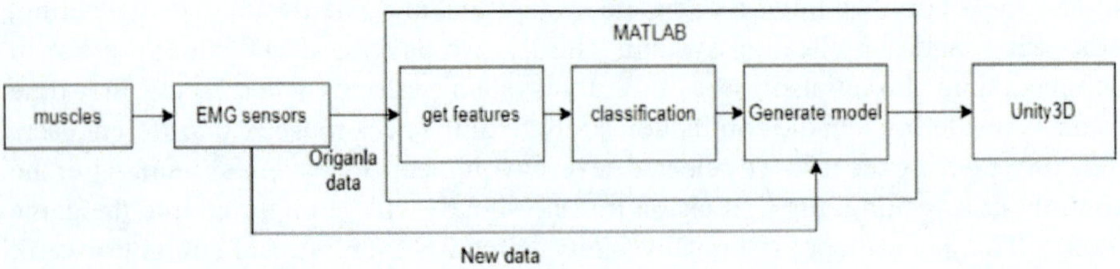

Fig. 1 The system of EMG-based view controller using VR applications

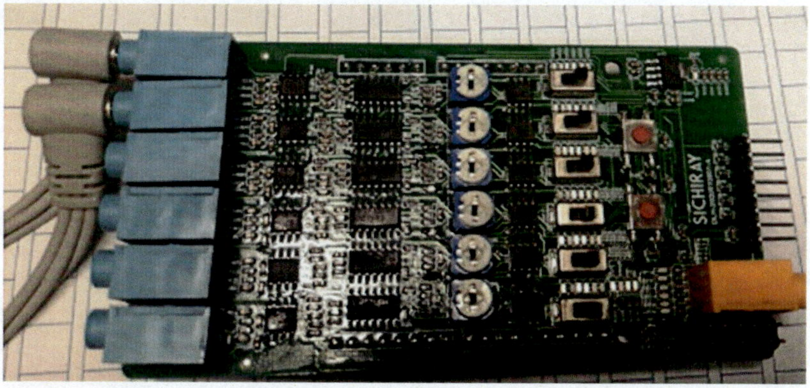

Fig. 2 Six-channel EMG data acquisition device

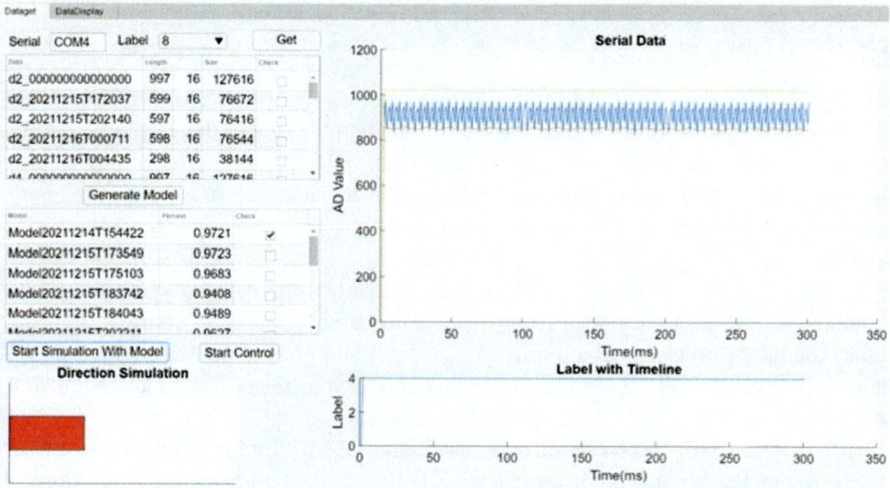

Fig. 3 MATLAB application developed platform to get serial data and do classification, model training, raw data management, and verify with new data

2.2 Data Process

The data sent by serial will be recorded by MATLAB. Most data processes will be done on MATLAB, including feature extraction, model training, and online working. A MATLAB application is also developed to control system's flow with a GUI interface. The GUI interface is shown in Fig. 3.

Fig. 4 Three-dimensional application to test output

2.3 Interface Software

Unity 3D is a game engine widely applied in mainstream game development. Unity 3D has a camera system to control the view in the 3D scene with input devices. In this case, the MATLAB output data, which is in the form of simulating keyboard pressing, can control the 3D camera to test and verify the performance of the EMG control system.

In the scene shown in Fig. 4, players could control the camera using the EMG method. Nine blocks are put in front of the camera and randomly changed into the red as the target. Players need to aim at the target and get points. Each round of the game will give players 30 seconds to get as many points as possible. The conclusion will take the score as one of the evaluation indexes.

EMG signals will be processed and classified with an average of latency less than 0.15 s under the testing environment of a laptop with Windows 10, Intel Core™ i7-8750H CPU @ 2.20GHz, 8GB RAM, and an NVIDIA graphics card GTX 1060, MATLAB2019a, and Unity 3D version 2020.3.22.

3 Methods

3.1 Feature Extraction

The testers of the experiment were asked to sit still and only move their heads according to the given direction. Their heads should be moving from the center to corresponding laws. One of the most critical requirements for the testers is turning their heads as much as possible until they cannot turn. Afterward, as the EMG signal is recorded and sent to MATLAB, a feature extraction function is then run and does the work of extracting features of the transmitted signal from each channel, including enhanced mean absolute value (EMAV), average amplitude change (AAC), waveform length (WL), maximum fractal length (MFL), and root mean square (RMS) [15]. The extracting process happens iteratively with an interval of 100 data points. In other words, the script happens to extract features for every 100 data points. Within 30 seconds of the recorded test data set, the script would generate continuous feature collections, represented as the matrix T, including three channels and five directions. The rightmost column inside the matrix T denotes the label of the intended directions, so that later in the classification process, the classifier could recognize the groups.

Figure 5 shows the raw data. The x-axis shows time; the y-axis shows analogue to digital value. We chose several 2D feature scatter plots in Fig. 6.

3.2 Classification

The data classification process is being activated when the real-time data is recorded. This supervised learning process ensures that EMG signals generated by a particular turn of the subject's head are classified as up, down, left, right, or center. Then, MATLAB will send the corresponding instructional commands to Unity to control the screen. To obtain a stable model for the data classification, the signals detected in the experiment are used to optimize the trained model. The analysis process adheres to the principle of using the same channel but different features, which leads to the comparison with the same channel with different features, EMAV, AAC, WLM, MFL, and RMS. It helps the classification learner application in MATLAB use SVM (support vector machine) to achieve the target accuracy of original data as high as 96.5% on average we tested. This model was stored and installed in the main application to support the data processing process.

4 Experiment

4.1 Sampling Muscles

4.1.1 About the Muscles Related to Experiments

In this experiment, we used nine electrodes, of which six electrodes were used to measure several EMG values of subjects. By consulting the literature and preliminary investigations, we determined the muscles needed for the investigation. The muscles measured by these six electrodes will have a blatant response (tighten or relax) when the subject moves his head and can get excellent and stable feedback simultaneously.

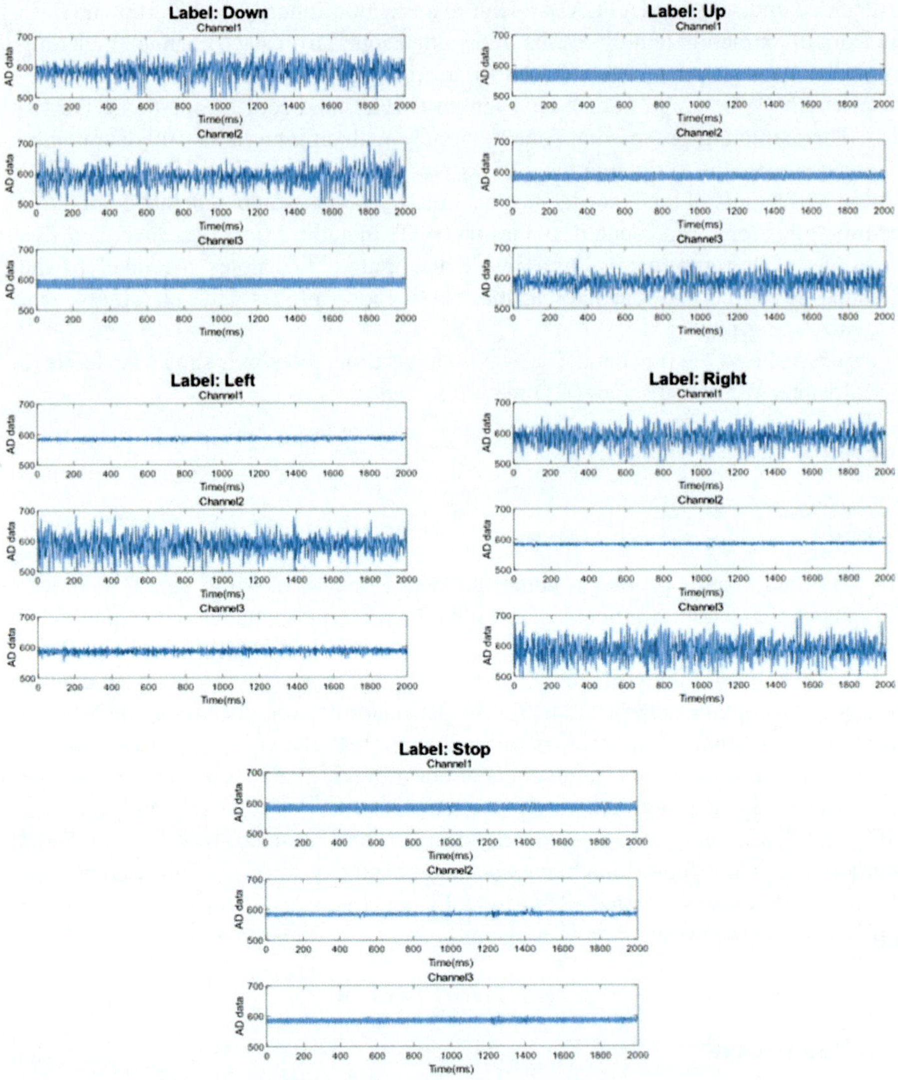

Fig. 5 The raw data from all labels shown above in three channels

In the following article, we use CH to represent the channel: R for red; G for green; Y for yellow.

In Fig. 7, CH1-Y and CH1-G are roughly located at the sternocleidomastoid muscle (SCM), of which CH1-G is closer to the sternum and clavicle. CH2-Y and CH2-G are located on the platysma muscle, and CH2-G is also more relative to the sternum and clavicle. The specific position is the position of the electrode in Figs. 7b and 8b when we are looking for a suitable muscle.

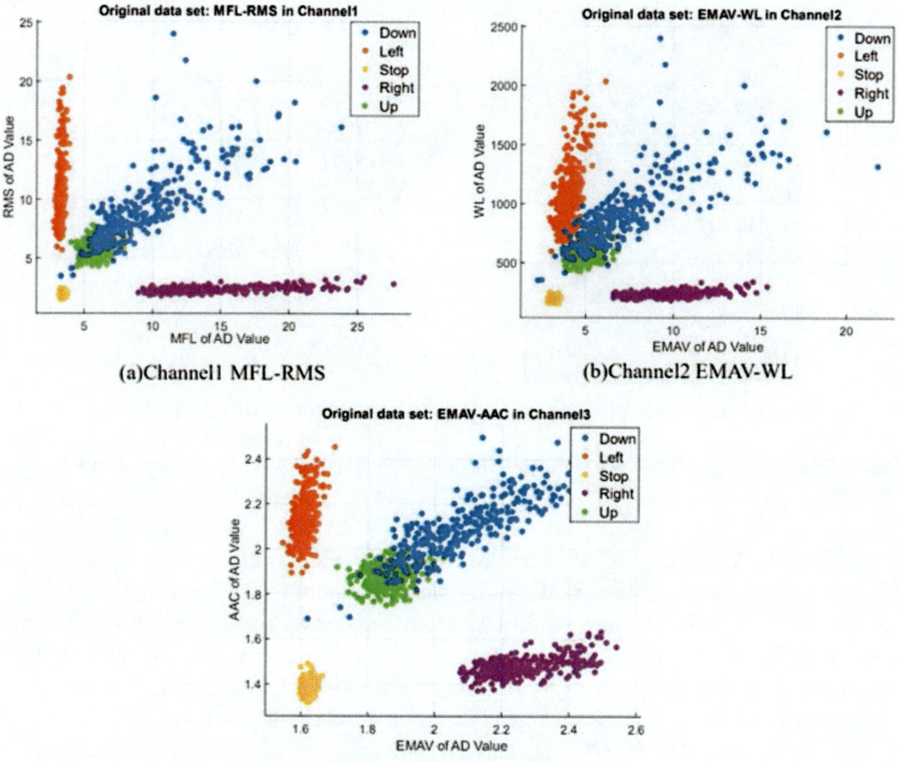

(a)Channel1 MFL-RMS (b)Channel2 EMAV-WL

(c)Channel3 EMAV-AAC

Fig. 6 The scatter plot of several specific features

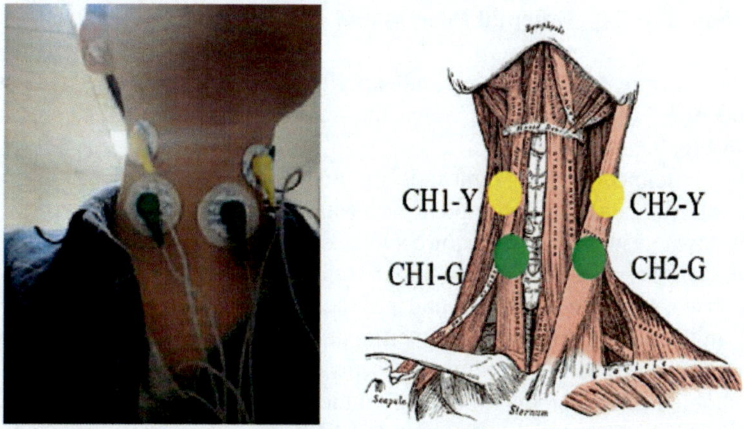

(a) Electrodes' positions (b) Atomy diagram for electrodes

Fig. 7 Front view for electrode settings. (**a**) Electrodes' positions. (**b**) Atomy diagram for electrodes

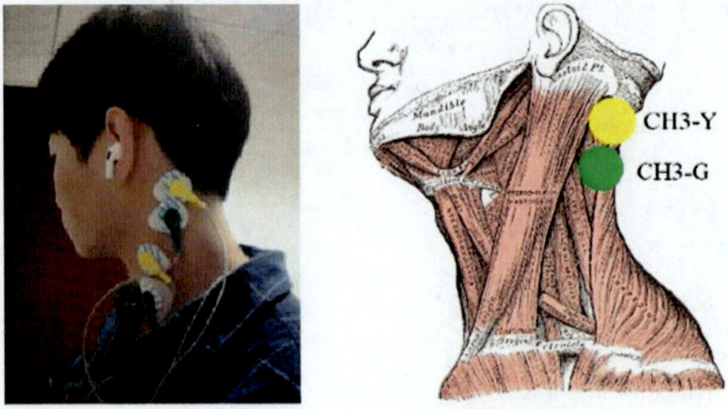

(a) Electrodes' positions (b) Atomydiagram for electrodes

Fig. 8 Side view for electrodes settings. (**a**) Electrodes' positions. (**b**) Atomy diagram for electrodes

As for CH3-Y and CH3-G, they are located at the splenius of our neck, behind our ears, which are shown in Fig. 8. We judge the movement of the head by measuring the electrical signals of these muscles when we move the head. Through our measurement, the muscles corresponding to A and B have an evident and stable response when we turn our head left and right. Their EMG signals enable the program to determine whether our heads are deflected to the left or right.

The electrodes CH1-R, CH2-R, and CH3-R are located at the junction of the sternum, and the cervical and thoracic vertebrae act as a grounding element in the measurement. This is what the remaining three electrodes do.

4.1.2 About the Experimental Process and Precautions

About a few gestures in our training and experiment part, we designed five gestures to measure the EMG signal and move the cursor position. The movements are showed in Fig. 9.

Labels are defined as stop, right, left, up, and down. The first four gestures are shown in (a), and the subjects complete each action by typical deflection of the head. The head-down gesture did not require subjects to lower their heads ultimately. As long as the subject makes a slightly larger action than the stopping range, the two muscles shown in (b) will tighten and produce a stable EMG signal that can be captured. In the training part of the experiment, we asked the subjects to keep each movement for more than 30 seconds when making gestures. At the same time, the subjects should achieve the maximum amplitude when turning the head and neck left and right so as to ensure that the generated EMG signal is most easily captured, transformed, and recognized by computer programs. At the same time, we will also provide subjects with a chair with a back to prevent them from leaning back when raising their heads; otherwise, they may cause different muscle reactions. After the

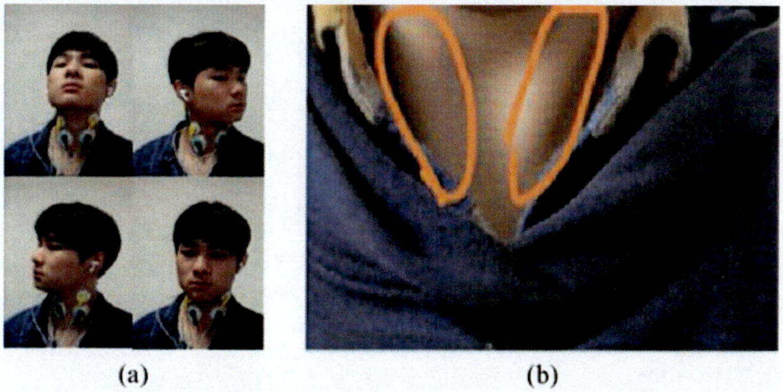

Fig. 9 (**a**) Schematic diagram of the head: up, left, right, and stopped movement. (**b**) The muscles in the neck that react when the head are lowered

training of each action, we let the subjects participate in a formal experiment, namely, a VR game. We completed this experiment by collecting the EMG signals of the subjects' muscles in the VR game.

4.2 Experiment Outcome

The latency between getting serial data from hardware and outputting by MATLAB is less than 150 ms per controlling instruction. According to the paper [16], a real-time EMG controller is acceptable having a latency than 300 ms, which shows that the system works at a stable status.

The experiment is designed to identify the feasibility of the view controller with the simulated data by comparing it to the traditional keyboard controller. To actualize the comparison, we made a 3D application mentioned above to simulate the situation where the testers' head can be imitated as a reticle. The targets on the vertical plane can be eliminated once the reticle gets them across. Through turning the head, the reticle can be mobile and directive due to contractions of some detected muscles. Faster head-turning speed results in higher responsiveness of the reticle, vice versa. The target on the vertical plane will emerge in a random place after the reticle has touched the previous one. In total, there is 30-second time in each round for testers to muscle control the reticle touching the targets, and five scores per target.

In comparison, testers are required to manually use a keyboard to play the same game. After completing three EMG rounds and keyboard rounds for testers, the total score is calculated by summing scores from three rounds and averaging them. The winner is the controller with the highest score in each tester. The experimental data is shown below (see Table 1).

Table 1 The game data we currently tested

Tester	Age	Gender	Model accuracy of original data (%)	Score (keyboard)	Score (EMG)
Tester 1	21	Male	96.4	131.67	78.33
Tester 2	16	Male	96.8	143.33	63.33
Tester 3	20	Female	95.7	121.67	75.00
Tester 4	20	Female	96.3	111.67	71.66

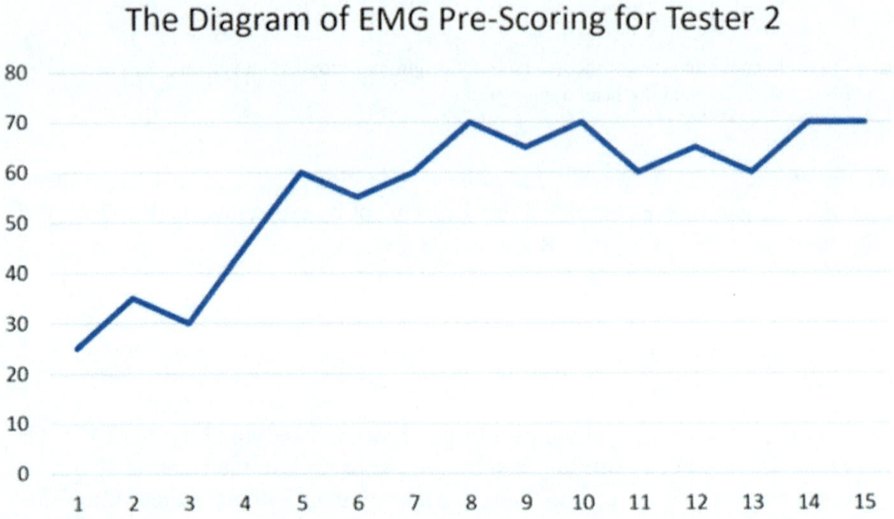

Fig. 10 The diagram of EMG pre-scoring for Tester 2

The keyboard controller takes the absolute advantage in terms of scoring, which signifies its superior responsiveness and accuracy compared to EMG. On the EMG side, it has been considered that testers need to get used to the new EMG controlling mechanism. So, we started the official game scoring after that. As the result of pre-scoring, we found that testers can achieve a relatively stable score, which is approximately within the range amid 60–70 scores, or 12–14 times for target elimination. It proves that testers can be accustomed to our EMG controller with a stable performance. On the other hand, compared to the keyboard, the testers reflected that turning the head requires more time to return and creates a higher possibility of misdirection. The diagrammatic EMG pre-scoring data has been shown below. Figure 10 shows the diagram of EMG pre-scoring for Tester 2.

5 Conclusion

We successfully developed a visual control device using the EMG method with the SVM model and mapping on the optical system and head orientation. We also showed that the EMG signal of the neck could effectively control VR applications with daily used movement. The system can perform an acceptable control model after acquiring 30 seconds per action which can be immediately used. The player can control the view by controlling muscles in VR application as convenient as in real life, which reached the goal of our system design. The accuracy, control effect, and latency are acceptable as the above data.

However, to realize more accurate control, we still need to do more algorithm work. The system can be advanced by adding angle regression to reach a more precise system with angle-formed output. In addition, latency also occurs on hardware preprocessing and serial-getting sessions, which can be optimized by an embedded program in further research.

References

1. B. Pandey, R.B. Mishra, An integrated intelligent computing model for the interpretation of EMG based neuromuscular diseases. Expert Syst. Appl. **36**(5), 9201–9213 (Jul. 2009). https://doi.org/10.1016/j.eswa.2008.12.023
2. T. Subba, T.S. Chingtham, A survey: EMG signal-based controller for human–computer interaction, in *Advances in Communication, Cloud, and Big Data*, (Singapore, 2019), pp. 117–125
3. M.B.I. Reaz, M.S. Hussain, F. Mohd-Yasin, Techniques of EMG signal analysis: Detection, processing, classification and applications. Biol. Proced. Online **8**(1), 11–35 (Dec. 2006). https://doi.org/10.1251/bpo115
4. J. Too, A. Rahim, N. Mohd, Classification of hand movements based on discrete wavelet transform and enhanced feature extraction. Int. J. Adv. Comput. Sci. Appl. **10**(6) (2019). https://doi.org/10.14569/IJACSA.2019.0100612
5. A. Phinyomark, P. Phukpattaranont, C. Limsakul, Feature reduction and selection for EMG signal classification. Expert Syst. Appl. **39**(8), 7420–7431 (Jun. 2012). https://doi.org/10.1016/j.eswa.2012.01.102
6. J.K. Muguro et al., Development of surface EMG game control interface for persons with upper limb functional impairments. Signals **2**(4), 834–851 (Nov. 2021). https://doi.org/10.3390/signals2040048
7. C.I. Christodoulou, C.S. Pattichis, Unsupervised pattern recognition for the classification of EMG signals. I.E.E.E. Trans. Biomed. Eng. **46**(2), 169–178 (Feb. 1999). https://doi.org/10.1109/10.740879
8. X. Liu, J. Sacks, M. Zhang, A.G. Richardson, T.H. Lucas, J. Van der Spiegel, The virtual trackpad: An electromyography-based, wireless, real-time, low-power, embedded hand-gesture-recognition system using an event-driven artificial neural network. IEEE Trans. Circuits Syst. II Express Briefs **64**(11), 1257–1261 (Nov. 2017). https://doi.org/10.1109/TCSII.2016.2635674
9. J. Kim, S. Mastnik, E. André, EMG-based hand gesture recognition for realtime biosignal interfacing, in *Proceedings of the 13th International Conference on Intelligent User Interfaces – IUI '08*, (Gran Canaria, 2008), p. 30. https://doi.org/10.1145/1378773.1378778

10. A.L. Rincon, H. Yamasaki, S. Shimoda, Design of a video game for rehabilitation using motion capture, EMG analysis and virtual reality, in *2016 International Conference on Electronics, Communications and Computers (CONIELECOMP)*, (Cholula, 2016), pp. 198–204. https://doi.org/10.1109/CONIELECOMP.2016.7438575
11. D. Blana, T. Kyriacou, J.M. Lambrecht, E.K. Chadwick, Feasibility of using combined EMG and kinematic signals for prosthesis control: A simulation study using a virtual reality environment. J. Electromyogr. Kinesiol. **29**, 21–27 (Aug. 2016). https://doi.org/10.1016/j.jelekin.2015.06.010
12. M.R. Williams, R.F. Kirsch, Evaluation of head orientation and neck muscle EMG signals as command inputs to a human–computer interface for individuals with high tetraplegia. IEEE Trans. Neural Syst. Rehabil. Eng. **16**(5), 485–496 (2008). https://doi.org/10.1109/TNSRE.2008.2006216
13. T. Sugiarto, C.-L. Hsu, C.-T. Sun, W.-C. Hsu, S.-H. Ye, K.-T. Lu, Surface EMG vs. high-density EMG: Tradeoff between performance and usability for head orientation prediction in VR application. IEEE Access **9**, 45418–45427 (2021). https://doi.org/10.1109/ACCESS.2021.3067030
14. Y. Barniv, M. Aguilar, E. Hasanbelliu, Using EMG to anticipate head motion for virtual-environment applications. I.E.E.E. Trans. Biomed. Eng. **52**(6), 1078–1093 (Jun. 2005). https://doi.org/10.1109/TBME.2005.848378
15. J. Too, Jx-EMGT : Electromyography (EMG) Feature Extraction Toolbox. 2022. Accessed: Feb. 09, 2022. [Online]. Available: https://github.com/JingweiToo/EMG-Feature-Extraction-Toolbox
16. U. Côté-Allard et al., Deep learning for electromyographic hand gesture signal classification using transfer learning. IEEE Trans. Neural Syst. Rehabil. Eng. **27**(4), 760–771 (Apr. 2019). https://doi.org/10.1109/TNSRE.2019.2896269

Comparison and Application of Two Face Detection Algorithms

Xiaoxi Wei and Qian Yin

Abstract This chapter compares two face detection algorithms Viola-Jones and MTCNN that have certain similarities in the structure of the algorithm. They are both creative and significant objective detection algorithms though they are based on different frameworks. To have a more intuitive view of the merits and demerits of two face detectors, we test two detectors on a challenging dataset including 25 images divided into 5 categories and compare the results in terms of detecting accuracy, time consumed on the detection, and effects of nonface features on detection results. Besides, for the results of MTCNN, it is essential to observe the positions of the five feature points which are key indicators of accuracy. By comparing the results, it is clear that the detection rate of MTCNN is generally higher than that of Viola-Jones, but the detection speed of Viola-Jones is faster. Furthermore, Viola-Jones has a good overall performance in positive face detection and is better suited for fast detection in frontal view, whereas MTCNN is better suited for industrial-grade scenes requiring high accuracy.

Keywords Face detection · Viola-Jones · MTCNN · Comparison

1 Introduction

The purpose of this paper is to learn the merits and demerits, differences, and connections between face detection algorithms Viola-Jones [1] and MTCNN [2] by comparing them and testing them on a challenging dataset.

X. Wei (✉)
School of Mechanical, Electrical and Information Engineering, Shandong University, Shandong, China
e-mail: 201900800198@mail.sdu.edu.cn

Q. Yin
School of Biological and Agricultural Engineering, Jilin University, Changchun, China

C.-C. Wang, A. Nallanathan (eds.), *Proceedings of the 5th International Conference on Signal Processing and Information Communications*, Signals and Communication Technology, https://doi.org/10.1007/978-3-031-13181-3_5

Viola-Jones is a machine learning-based object detection algorithm. MTCNN is a deep-cascaded multitask framework that takes advantage of the inherent correlation between them to improve their performance.

Both Viola-Jones and MTCNN are cascade face detectors. From the paper [2], it can be known that the proposal of MTCNN is partly inspired by the Viola-Jones algorithm. The first algorithm trains multiple cascading classifiers to replace a single classifier, whereas the second algorithm employs a cascaded structure with three carefully designed networks.

The Viola-Jones framework is one of the most landmark achievements in the history of face detection, laying the foundation for the AdaBoost-based objective detection framework, while MTCNN is a representative of algorithms based on deep learning framework. So these two algorithms which have certain similarities in the structure but are based on different frameworks are chosen for comparison to delve into the differences in detecting results.

Although there are more than 10 years between the Viola-Jones detector and MTCNN, MTCNN takes both the advantages and shortcomings of Viola-Jones into account. MTCNN obviously improves the accuracy of face detection. The application characteristics of the two algorithms, as well as their advantages and disadvantages, can be generally revealed by testing the two algorithms on the challenging dataset.

With the outcomes it is discovered that the Viola-Jones algorithm is more for the frontal face. For the complete face, it has a high recognition rate. But for different poses or a variety of shields, it will be difficult. Although MTCNN can recognize faces in different poses and even occlusion well, it sometimes fails to recognize some obvious faces among multiple faces. To explore more, we select dozens of images on the Internet and divide them into five categories. Although the number is small, they all have their own characteristics and can bring challenges.

The main issues addressed in this report are as follows: (a) Find 25 more challenging images as a small dataset to test each of the two algorithms and compare the test results. (b) Pick 2–3 ROI regions in each image, test them, and compare the results. (c) Summarize the results of the two algorithms, their strengths and weaknesses, and the scenarios in which algorithms can be used.

2 Related Work

One of the contributions of the Viola-Jones algorithm is using integral images for fast computation of Haar-like features, which is motivated in part by the work of Papageorgiou et al. [3]. And the AdaBoost method proposed by Yoav Freund et al. [4] is used to build the classifier.

With the emergence of convolutional neural networks (CNNs), some people use them in a range of computer vision applications and make lots of impressive advances [5, 6]. Some CNN-based face detection algorithms were presented, inspired by the good performance of CNNs in computer vision tasks. Yang et al.

[7] build deep convolution neural networks for facial attribute detection to achieve high response in face areas, yielding candidate windows of faces. This technique, however, is very consuming in practice due to its sophisticated CNN structure. Li et al. [8] utilize cascaded CNNs for face detection. However, it has some limitations in terms of application, incurring additional computational costs and ignoring the intrinsic connection between facial landmark localization and bounding box regression. For the image pyramid constructed in the first step of MTCNN, the candidate window and its boundary box regression vector are obtained by a method similar to the Deep Dense Face Detector (DDFD) [9].

2.1 Viola-Jones

The overall flow of the algorithm can be summed up as follows:

- Application of Haar-like input features: thresholding the sum or difference of rectangular image regions.
- The new image representation technique speeds up the computation of values for a 45-degree rotation of the rectangular image region, and this image structure is employed in order to accelerate the computation of Haar-like input features.
- Using AdaBoost to create classifier nodes for the binary classification problem (high pass rate, low rejection rate) (face vs. nonface).
- Construct a filtered cascade of classifier nodes (in a filtered cascade, a node is a set of classifiers of the AdaBoost type). To put it in another way, the first set of classifiers is optimal and can pass through the region of images containing objects while allowing some images without objects to pass; the second set of classifiers is suboptimal and has a low rejection rate.

Viola-Jones achieves good performance while remaining real-time efficient. However, several studies [10–12] suggest that the performance of this detector may deteriorate significantly in practical applications with highly variable face morphology, even when using more advanced features and classifiers [2].

2.2 MTCNN

The mainframe of MTCNN is like a cascade with three neural network stages. These include proposal network (P-Net), refine network (R-Net), and output network (O-Net). The features of MTCNN come from each image that it is detecting. It can accurately identify face images with different poses and different occlusion.

MTCNN transforms the image to form an image pyramid and to detect faces of different sizes. The results are entered into P-Net for preliminary testing. The features of MTCNN are all learned from the image itself.

P-Net is a full convolution network. For the image pyramid constructed in the previous step, the candidate window and its boundary box regression vector are obtained by a method similar to Deep Dense Face Detector (DDFD). After the features of this network are input into three convolution layers, a preliminary face classifier is used to briefly determine whether this region is a face or not.

The second layer is called R-Net, and comparing with the first layer P-Net, a full connection layer is finally added. The candidate Windows screened by P-Net are input into this layer, and a large number of poor candidate Windows are filtered out. Finally, the results are further optimized by boundary box regression and NMS (non-maximum suppression) to output more reliable face results.

The last layer structure, O-Net, adds another convolution layer to R-Net. O-Net is a more complex convolutional network that has more input features and also has better performance, with the output of this layer as the final network model output. Similar to R-Net, both use a more complex structure for better optimization.

As shown above, both algorithms are very creative in their formulation, and their structures have similarities and differences. To better understand the strengths and weaknesses of these two algorithms, as well as the scenarios in which they are used, we select 25 more challenging images to test and compare the results of the two algorithms.

3 Experimental Setup

In this section, we write complete and working code based on the two algorithms and compile and test the code in a compiler to ensure that the code can run properly to obtain face detection results. We then select 25 images from the web to constitute a representative dataset for testing. To make the results more comprehensive and comparable, the images are divided into five categories with five images in each:

- Pose: This category contains images of people looking in various directions: side, top, and bottom. The objective is to evaluate the detector's accuracy in detecting faces at different viewing angles.
- Partial occlusion: In the presence of occlusion, the accuracy of face detection and face landmarks detection is reduced. There are five occlusion-affected images to investigate the extent to which occlusion affects the two face detectors.
- False pose: Images that resemble a human face can be detected as faces in some cases. Five images of monkeys, an avatar, a smiling face, and a dog are gathered to see if the two face detectors would recognize them as faces.
- Age: Given that some infants' facial features are not fully mature and the elderly are not as distinguishable as the young due to their age, five photos of infants, children, and elderly people are found to see if they could all be detected.
- Races: Images of people with various skin tones are selected to test the effect of skin tone on the two face detectors.

All of the images are input into the detector for testing, collecting the results, and comparing the results of the two different algorithms. To make the comparison more intuitive, the results are evaluated from three aspects: precision, recall, and FPS (frames per second). These metrics provide a more comprehensive assessment of the face detector's results. After obtaining these results, a more overall view of the performance of the two detectors can be obtained. As a result, it is more accessible to recognizing the strengths and weaknesses of each of the two algorithms, as well as the various scenarios in which they are applicable.

4 Experimental Results

The results are obtained by using two face detectors on each of the 25 images in this section. To visualize these metrics, we calculate the precision and recall for all images and test the time of detection for each image. The results are shown in Table 1 and Table 2.

4.1 The Results of Viola-Jones

The results show that the Viola-Jones algorithm has a rather low side-face detection accuracy but a high front-face detection accuracy, while age and different races (skin color) have little effect on it, and the majority of images in these two categories are detected with high accuracy. As can be seen, the precision of "age"' reaches 100%. However, occlusion is a problem for the Viola-Jones detector, as evidenced by the fact that detection accuracy for images with occluded faces is as low as 38%, with

Table 1 The evaluation results of Viola-Jones

Category	Precision (%)	Recall (%)	FPS
Pose	15	10	23.85
Partial occlusion	38	36.36	21.71
False pose	40	40	20.11
Age	100	100	11.89
Different races	74.23	85.81	15.37
Average	53.45	54.43	18.58

Table 2 The evaluation results of MTCNN

Category	Precision	Recall	FPS
Pose	100	100	10.64
Partial occlusion	98.18	98.18	5.72
False pose	40	40	5.14
Age	100	100	1.85
Different races	79.55	91.91	1.89
Average	84.55	86.02	5.05

Fig. 1 The results of Viola-Jones. The image on the left is the only image that is detected in its entirety in "partial occlusion"; on the right, an area on the camouflage is misidentified as a human face. (**a**) Result 1. (**b**) Result 2

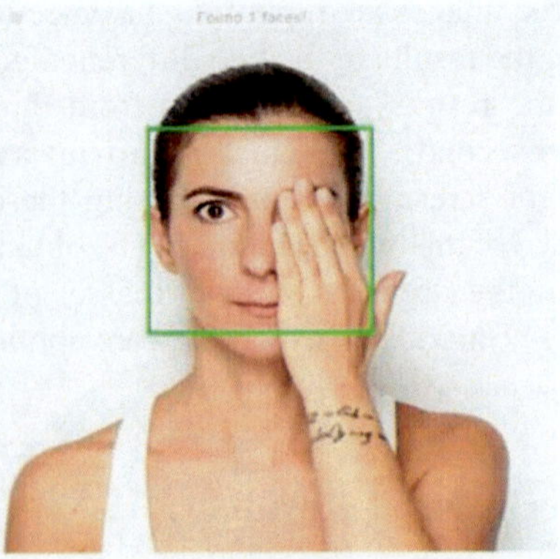

(a) Result 1

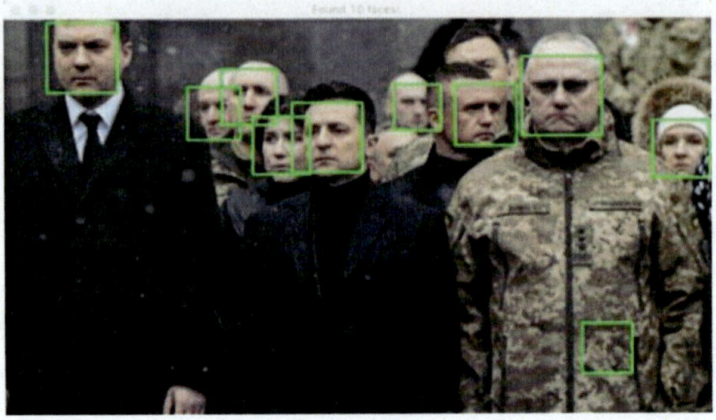

(b) Result 2

only one image in this category having an accurately detected face (as Fig. 1, Result 1). It is funny to see that the camouflage in one image is mistakenly detected as a face (Fig. 1, Result 2), so we select several more images of people wearing camouflage for detection. The detector is still disturbed by the camouflage, which proved to have a relatively large effect on the Viola-Jones detector.

There is also a type of image that resembles a human face. Images of monkeys and dogs in this category can be misidentified as human faces by Viola-Jones detector. We look for other pictures of dogs and monkeys for this purpose and discover that only pictures of animals with features that are very similar to human faces can be detected as human faces, other pictures cannot (Fig. 2, Result 2).

It is also possible to determine that the Viola-Jones is better at detecting frontal faces and faces with low occlusions by looking for commonalities in images with good detection results.

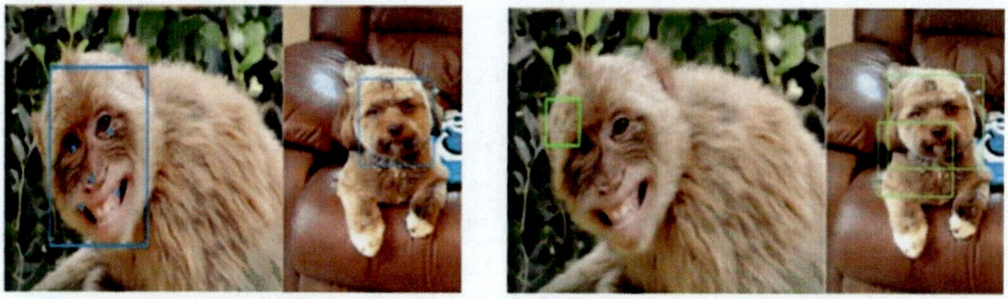

(a) Result 1 (b) Result 2

Fig. 2 The monkey and dog in MTCNN and Viola-Jones. The left two pictures are results of MTCNN and the right two are from Viola-Jones. It is interesting to find out that both algorithms have trouble dealing with monkeys and dogs. (**a**) Result 1. (**b**) Result 2

Fig. 3 Example of age. This picture is from the age images, it is clear that different ages do not make MTCNN confused. So does the partial occlusion

4.2 The Results of MTCNN

What is pleasantly surprising is that MTCNN recognizes faces in all poses, ages, and races with extremely high accuracy across all five image categories—just like Fig. 3. However, for some multi-face images, a face cannot be recognized but misjudged clothes. In addition, MTCNN is also low on false face recognition, recognizing monkey and dog faces as faces.

Specifically, for the identification of age categories, the accuracy reaches 100%. The same is true for poses (side face). But for different races, producing a face without recognition is only 79.55% accurate. For partial occlusion, the accuracy is only 98.18%. The results are shown in the Table 2. In the judgment of the "false pose," the correct recognition rate is as low as 40%. Especially for monkey and dog faces (Fig. 2, Result 1), the error rate is high. In subsequent tests with monkeys and dogs, the error rate is 66.7% for monkeys and 40% for dogs. We use different images of monkeys and dogs to further test algorithms. Along with the one depicted in

Fig. 4 Different faces with two results. The face of the monkey on the left is not detected as a face, whereas the face of the monkey on the second right is recognized as a face by the detector

Fig. 4, it is found that with a black face and a white circle around it, the left monkey is a true negative picture. However, the right picture is false negative. The reason of this may be the monkey face color. The left monkey is far away from a human being, but the right one is more similar to a human face.

When the data in the table is compared, it is clear that MTCNN has a higher precision and recall than Viola-Jones, implying that the first detector has a higher overall accuracy. However, when we extract 2–3 ROI regions from the images, such as undetected faces or regions misidentified as faces, all of the results are promising. Furthermore, the data in the table's last column shows that the average FPS (frames per second) of Viola-Jones is 18.58, while the MTCNN detects images at 5.05 frames per second. This suggests that in terms of detection speed, the Viola-Jones outperforms the MTCNN.

Based on the findings, a conclusion could be drawn. It appears that the Viola-Jones detector is significantly faster than the MTCNN detector. Both two detectors do poorly against animals such as monkeys and dogs (Fig. 2). And both of them are prone to misjudgments about certain parts of camouflage.

Analyzing the test set revealed that the algorithms have errors in animals due to hair occlusion. But based on subsequent tests and further analysis, the hair, while likely to have an impact, is not as big as expected. It is more likely that the algorithm does not look at the color of the site when detecting whether it is a face. If an animal's face is the same color, it may misjudge and recognize it as an adult face (Fig. 4). The same is true of camouflage clothing, which will be misjudged if there are patches of color that resemble a human face.

5 Conclusion

This paper compares the two object detectors of Viola-Jones and MTCNN. They are tested in terms of detection speed and accuracy on different types of images. According to the experimental results, although Viola-Jones is quick to test faces, only the frontal face has relatively reliable accuracy, and the other poses cannot be applied widely in practice. MTCNN goes a long way in this respect, as it can reliably test faces in different poses. But at the same time, MTCNN's program execution in this experiment depends on the image itself and how much time it takes.

For MTCNN, it is more used for testing faces in different poses and requires high accuracy, but for scenes that take less time to consider. Especially when images containing animals cannot be examined, errors can easily occur. For Viola-Jones, the application scope is narrower and can only be applied to the detection of front human face images, which takes less time and has low relative accuracy. But the same cannot be said for scenes that contain images of animals.

When comparing results, it is impossible to take into account all of the influences. What is certain is that MTCNN outperforms Viola-Jones in terms of detection accuracy and is less vulnerable to occlusion and other factors. However, Viola-Jones's overall detection rate is much faster than MTCNN's.

However, both algorithms fail to recognize the images that are very similar to human faces. They only detect whether it is a face or not based on a few features, but do not strictly distinguish between humans and other species or objects, which is most likely why, in some instances, animals or clothing are also detected as faces as well.

The algorithm could be improved in a variety of ways through the analysis. The underlying causes of why some animals (such as monkeys and dogs) are prone to facial misjudgment can be investigated. It is also potential to increase the precision of the results by expanding the training set and using a detector to separate clothes and other background from the face before performing face detection—in particular, from the aspect of facial color, differences between human faces and other objects, to better improve the performance of the detector.

References

1. P. Viola, M.J. Jones, Robust real-time face detection. Int. J. Comput. Vis. **57**(2), 137–154 (2004)
2. K. Zhang, Z. Zhang, Z. Li, Y. Qiao, Joint face detection and alignment using multitask cascaded convolutional networks. IEEE Signal Proc. Lett. **23**(10), 1499–1503 (2016)
3. M.T. Pham, Y. Gao, V.D.D. Hoang, T.J. Cham, Fast polygonal integration and its application in extending haar-like features to improve object detection, in *2010 IEEE Computer Society Conference on Computer Vision and Pattern Recognition*, (2010, June), p. 942949
4. Q. Zhu, M.C. Yeh, K.T. Cheng, S. Avidan, Fast human detection using a cascade of histograms of oriented gradients, in *2006 IEEE Computer Society Conference on Computer Vision and Pattern Recognition (CVPR'06)*, vol. 2, (2006, June), pp. 1491–1498
5. C.P. Papageorgiou, M. Oren, T. Poggio, A general framework for object detection, in *Sixth International Conference on Computer Vision (IEEE Cat. No. 98CH36271)*, (1998, January), pp. 555–562
6. Y. Freund, R.E. Schapire, A decision- theoretic generalization of on-line learning and an application to boosting. J. Comput. Syst. Sci. **55**(1), 119–139 (1997)
7. A. Krizhevsky, I. Sutskever, G.E. Hinton, Imagenet classification with deep convolutional neural networks. Adv. Neural Inf. Proces. Syst. **25** (2012)
8. Y. Sun, Y. Chen, X. Wang, X. Tang, Deep learning face representation by joint identification-verification. Adv. Neural Inf. Proces. Syst. **27** (2014)

9. B. Yang, J. Yan, Z. Lei, S.Z. Li, Aggregate channel features for multi-view face detection, in *IEEE International Joint Conference on Biometrics*, (2014), pp. 1–8
10. S. Yang, P. Luo, C.C. Loy, X. Tang, From facial parts responses to face detection: A deep learning approach, in *Proceedings of the IEEE International Conference on Computer Vision*, (2015), pp. 3676–3684
11. H. Li, Z. Lin, X. Shen, J. Brandt, G. Hua, A convolutional neural network cascade for face detection, in *Proceedings of the IEEE Conference on Computer Vision and Pattern Recognition*, (2015), pp. 5325–5334
12. S.S. Farfade, M.J. Saberian, L.J. Li, Multi-view face detection using deep convolutional neural networks, in *Proceedings of the 5th ACM on International Conference on Multimedia Retrieval*, (2015), p. 643650

Research on Sound Field Evaluation of Urban Walking Street Through Sound Environment Perception

Wei Lin and Yi-Ming Wu

Abstract Recreation space plays an important role in the rapid development and evolution of the city, and using visual perception theory to explore the connection between social history, cultural significance, and environment has attracted increasing attention. In recent years, the study of urban soundscape has shifted the focus from the typical urban image and self-evident argument to neighborhood sound field characteristics of walking street space, to explore the perception of the overall urban environment and situational shaping. The description and psychological evaluation of spatial soundscapes are multidimensional and interactive. The primary consideration of the physical quantity of sound energy is an important reference for the design and planning of walking street sound field. The research takes the footpath system adjacent to urban walking space as the object and discusses the sound energy distribution of its sound field from multiple dimensions such as time, energy, space, and perception. In this study, the objective physical measurement results of LAeq (dB) are used as the basic prototype of the sound field, and regional and sound energy distribution maps are gradually expanded with electric model simulation software drawing point, line, and plane. The relationship between sound environment perception and spatial visibility of Calligraphy Greenway recreation space is explored by analyzing the spatial visibility and combining the perception of the sound environment and spatial visibility. The study mainly discusses the general characteristics of the sound field distribution and the main sound sources of Calligraphy Greenway recreation space and investigates the difference of spatial sound environment caused by non-acoustic factors based on the street's spatial pattern. The research results will provide theoretical support for urban design optimization and serve as a decision-making possibility for the development of tourism and leisure space in Taiwan.

W. Lin
School of Architecture, Feng Chia University, Taichung, Taiwan, China

Y.-M. Wu (✉)
Ph.D. Program for Civil Engineering, Water Resources Engineering, and Infrastructure Planning, Feng Chia University, Taichung, Taiwan, China

C.-C. Wang, A. Nallanathan (eds.), *Proceedings of the 5th International Conference on Signal Processing and Information Communications*, Signals and Communication Technology, https://doi.org/10.1007/978-3-031-13181-3_6

Keywords Recreation space · Objective indicators · Sound field simulation · Soundscape perception

1 Introduction

Walking is the main form of human movement, which is a necessary means for experiencing and enjoying urban space. Therefore, walking streets are regarded as a necessary public space, and the sound environment of walking streets directly affects the living feeling and comfort of urban residents. Unlike the dense road net in other places, the road net spacing in Taiwan is much smaller, with more motorcycles. One third of its main roads have a traffic noise of over 70 dBA, and many walking streets are adjacent to main roads, with serious noise pollution and poor sound comfort. In recent years, among the relevant economic activity data or social science research, most international environmental policies focus on noise control and have gained considerable research achievements in energy-time domain and spectrum domain, which can reduce noise level though does not necessarily improve the quality of urban life [1, 2, 3, 4]. The World Health Organization and various national indicators are trying to optimize the high-decibel sound energy generated by transportation and industrial production, to meet regional specifications (World Health Organization, 1999 [5]; European Parliament and Council, 2002 [6]; World Health Organization, 2011 [7]), especially the urbanization development index, so as to actively create a high-quality living environment of low voice energy, which has become an important reference for livable cities. Relevant research also shows that lowering sound levels does not necessarily lead to a better sound environment in urban areas [8]. These measures should be extended to soundscape construction [9]. Therefore, it is of great significance to analyze the sound environment perception of walking streets. Research on the influence of sound environment perception has been carried out at home and abroad. For example, Kang Jian et al. [10] discussed the subjective evaluation of urban noise, and Zhang Yuan et al. [11] proposed to improve sound environment perception in public space from the perspective of soundscape restoration. However, these studies mostly discuss sound environment from the perspective of public space, ignoring the study of the sound environment of areas with a relatively bad sound environment such as the walking street near the main road. Ye Yu et al. [12] proposed an operational framework for measuring the space quality of human-scale-oriented streets, but the research focuses more on the spatial quality or visual quality of living streets, ignoring the sound environment perception of walking streets. In a walking street, people's perception of the sound environment not only depends on the size of sound energy but also has a lot to do with pedestrians' subjective perception. Urban spatial patterns, such as functional formats and spatial visual quality, will also affect pedestrians' perception of the sound environment. This research measured and investigated the sound field status of urban recreation space and obtained the sound environment and sound propagation law in streets through computer simulation of building sound environment. The research results can be used to measure whether sound energy is fuzzy noise or to classify and define frequency characteristics. What's more, it can also provide the soundscape perception and analysis of the next stage, which will be the main follow-up work of this study.

2 Spatial Visibility Theory and Spatial Pattern of Walking Streets

2.1 Spatial Visibility Theory

Spatial visibility is an investigation of the spatial visual form of the study site, including green visibility, enclosed pattern, sky visibility, and degree of motorization. These four indicators were acquired with the following method: taking panoramic photos of the research site with mobile phones and segmenting these street view images intelligently with SegNET, a pixel-level semantic segmentation system, to identify 12 categories of elements in the street: plants, sky, buildings, footpaths, pavement, fences, cars, telephone poles, pedestrians, road signs, bicycles and signs, as shown in Fig. 1. The specific visual form elements include the four indicators, as shown in Fig. 1.

Green visibility is the proportion of plants; sky visibility is the proportion of sky; the enclosed pattern is the proportion of buildings, columns, fences, and plants; and the degree of motorization is the proportion of roads and cars. Previous studies have shown that the relationship between sound environment perception and spatial visibility of walking streets generates obvious clusters. Cluster 1 is an open-space type, which has relatively open space and a complex functional format. Cluster 2 is an enclosed type, which has relatively compact space and a simple functional format. The sound environment perception cluster is closely related to spatial visibility indicators. The open-space type is more sensitive to changes in sound perception caused by changes in spatial visibility indicators than the enclosed type [13]. As walking streets next to main roads change from open type to enclosed type, with the functional format changing from complex to simple, the degree of acoustic comfort and harmony gradually increases, and loudness gradually decreases [14]. Calligraphy Greenway recreation space belongs to the open type, so the change of sound perception caused by the change of spatial visibility indicators is relatively sensitive.

2.2 Spatial Pattern of Walking Streets

The study is to research the sound environment in the space of Calligraphy Greenway, taking the greenway space from the Museum of Natural Science to Civil Square as the research object, as shown in Fig. 2. The common noise source in the street is the main sound source. By analyzing the main influencing factors of the street sound environment and sound propagation in urban life, the sound environment of this

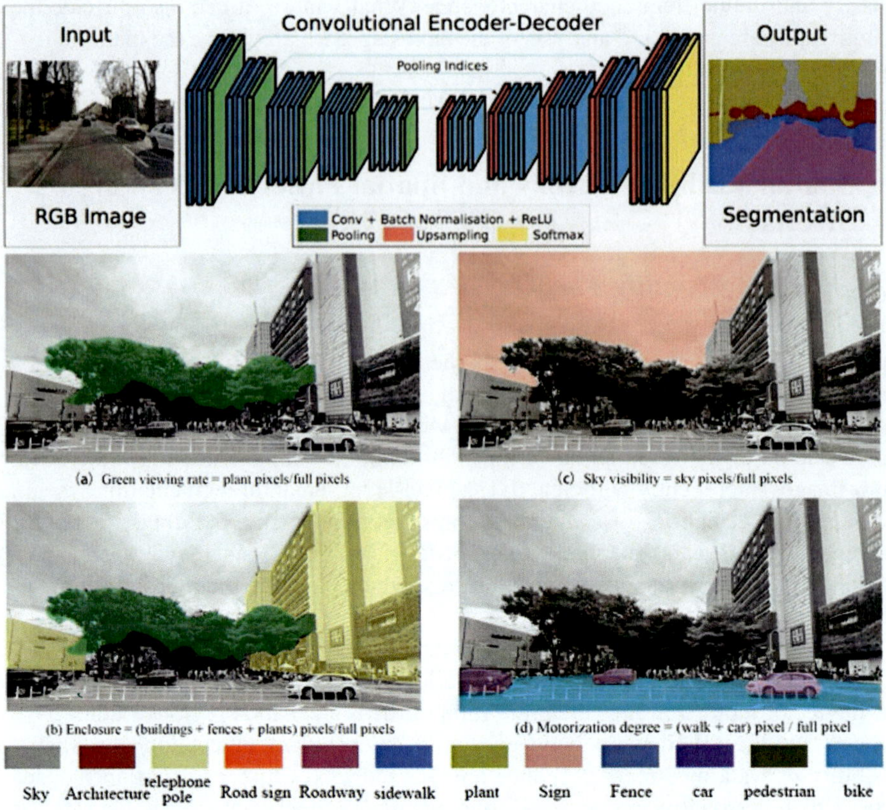

Fig. 1 Visual image analysis

living street was measured and analyzed from three aspects: street noise source, street spatial pattern, and street interface form. The improvement of landscape hardware facilities, the changes of the natural landscape, and the planting design configuration of this recreational walking street contribute to the unique space imagery atmosphere of Calligraphy Greenway. In addition, it introduces many art and cultural elements such as thematic installation art exhibitions, street performers, and creative markets, making it gradually full of spatial charm and urban cultural characteristics. Calligraphy Greenway aims to create a green garden road connecting Taichung sightseeing resources, and a unique outdoor exhibition hall carrying local culture, and to build it into an international sightseeing base. However, although the sound field of the walking street is attractive, the main key indicators that affect it need to be clarified. At present, according to survey results, the characteristics of Calligraphy Greenway soundscape can be preliminarily classified. The sound field of walking space along the street in the commercial area and market can be investigated through sound measurement and recording, the data of which would become the important basic data of soundscape perception. At the present stage, the

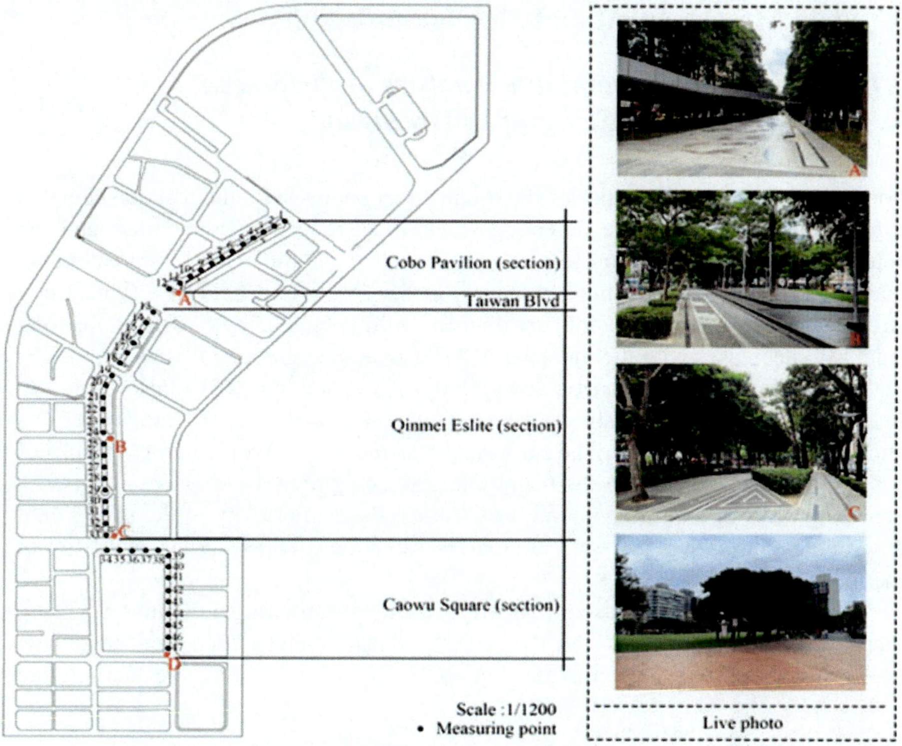

Fig. 2 A schematic diagram of the distribution of each measuring point

measurement of Calligraphy Greenway sound energy can provide a complete basis for sound field distribution. Calligraphy Greenway is about 1000 m long, with a one-way road of 10 m on both sides and a walking and the recreational road of 40 m. The height ratio between the building and the street is about 1:0.3 ~ 1:1.5. The street scale is unique. The measurement was carried out on weekdays and weekends, and a total of 47 groups of measurement points were evaluated and recorded by LAeq (dB) (measurement was conducted at 47 measuring points at 10:00, 15:00, and 22:00, respectively). The measuring time is 1 minute for each point. Through this way as mentioned above, the LAeq(dB) value between each measuring point of Cao Wu Dao in different time periods is analyzed. (Note: LAeq(dB) value would be influenced whether the measurement time is too long or too short. Too long, LAeq (dB) value would be excessively low; too short, LAeq(dB) value would be excessively high. It was concluded after comparing the results of measurements at a different time that 1 minute would be perfect to measure LAeq(dB) value and the result would be representative.) The objective evaluation of the soundscape was obtained, and the information would become the important basic data for the discussion of soundscape perception in the next stage. The field measurement distribution and photo schematic diagram of each measurement point are shown in Fig. 2.

3 Field Measurement and Simulation Results

3.1 Objective Evaluation and Analysis of the Sound Environment of Calligraphy Greenway

For the physical characteristics of the urban street sound field, the residual sound is usually much shorter at the reflection and diffuse reflection boundaries, and the sound attenuation at the distance between the source and the receiver is greater than that at the geometric reflection boundary. In the free sound field of a typical city street, when the length and width are doubled, SPL decreases by 6–9 dB in the long side; SPL decreases with the increase of distance and height, and it is usually 8 dB; with the increase of the acoustic absorptivity of the boundary, SPL also decreases linearly, and when the acoustic absorptivity increases from 0.1 to 0.9, the acoustical attenuation is usually 12 dB. The study analyzed the Leq (A) measurement results of sound field characteristics at the main points in Calligraphy Greenway, and discovered new ways of exploring sound, and learned the particularity of the sound field environment, as shown in Fig. 3. It also described the sound field characteristics of the main points.

C1 is in front of the museum square (Taichung Museum of Natural Science), where the main sound sources are the sound of water, conversation, wind, and a little traffic sound; C2 is in the middle section of Boulevard, where the main sound

Fig. 3 Survey of CaoWu Trail Walk Survey

sources are the sound of conversation and leaves rustling; C3 is in the intersection of Taiwan Avenue and Guanqian Road, close to Taiwan Avenue, with the main sound source being the sound of conversation; C4 is around Taiwan Avenue and Zhongxing Street, close to Taiwan Avenue, and the main sound source is the large traffic flow; C5 and C6 are around Zhongxing Street and Yingcai Road, with the main sound source being conversation and leaves rustling and traffic sound occasionally; C7's terrain is relatively flat and is often used for various events, and the sound source is mainly the sound of conversation, advertising, and music; C8 has many steps and can provide people with leisure and entertainment, so the sound source here is mostly advertising, music, and conversation; C9 is in the square before Eslite Green Garden, where many people gather during holidays, so the sound source is mainly the sound of conversation, advertising, and music; C10 is in Calligraphy Greenway Square, where occasionally holds large-scale activities, and it is also a popular place for people to relax during holidays for it happens to be close to Gongyi Road, and the sound source is mainly the sound of conversation, advertising, music, and traffic.

According to the above figure and average field measurement results, the different time nodes and sound field distribution of the main walking recreation space of Calligraphy Greenway were discussed. Forty-seven measurement points along the walking street were selected, and the sounds were recorded at different fixed times and locations. The above analysis shows that Calligraphy Greenway is mainly composed of natural sound and man-made sound, which influence each other. All sound sources in this area can be divided into four categories: (1) natural sound; (2) equipment sound; (3) traffic sound: car sound and music; and (4) man-made sound: footsteps, playfulness and talking, etc.

As mentioned above, the LAeq(dB) value between each measuring point of Calligraphy Greenway in different time periods is analyzed. (Note: Calligraphy greenway is a kind of urban green belt walkways designed to meet recreational demands. The study mainly recorded the LAeq(dB) value of calligraphy greenway when there are people walking or after people leave. It was found that on weekends, people usually come to the greenway at 10:00 am and the LAeq(dB) value would reach its peak at 15:00. At 22:00, after people leave, the LAeq(dB) value of the greenway would fall to its normal value.)The measuring results are shown in Fig. 4. The noise mainly occurs between 10:00 am and 5:00 pm, and there is little noise at night. For the main road, the maximum sound energy can reach 80 dB (A), with the daytime average being 69.5 dB (A) and night average being 69.5 dB (A), a little decline. The results show that the noise is mainly concentrated between C7 and C9 (point 25 ~ point 33), and the sound field energy begins to decrease steadily at C10 both during the daytime and at night.

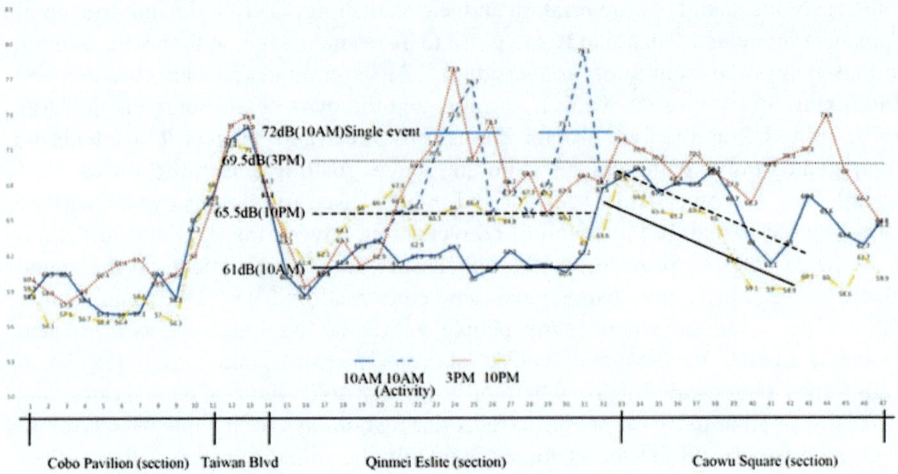

Fig. 4 Analysis of measurement points in the recreational space of Caowu Trail

3.2 Computer Simulation of the Sound Environment of Calligraphy Greenway

A sound energy distribution map was drawn based on the actual measurements, and the German NoiseAtWork software was used to analyze the map and draw the distribution curve. The results can be used to establish the optimal planning of the overall distribution of sound field, and individual environmental factors can be used for verification and hypothesis to construct models. Through the corresponding relation of measurements, the SPL image was drawn and the sound field model was verified, to predict future scenarios and benchmarks for subjective evaluation, and then take measures. Based on the results of the sound field environment generated by computer simulation, the grid calculation method was used to draw the curve of sound energy distribution, which has an intuitive trend and visual presentation and can evaluate the sound field environment in different conditions. The simulation results provide a lot of information, including the location and intensity of sound energy. In some cases, this method can be used for trend prediction, especially for stable sound sources, such as fixed noise sources (fixed mechanical sound). The measurement results of multiple sound sources can be discussed and evaluated simultaneously. For different sound sources, measures to reduce sound energy can also be mapped and ranked according to the size of sound energy results. NoiseAtWork was used to simulate and draw the sound energy distribution curve, so as to establish a framework of the overall distribution of the sound field. Individual environmental factors can be used for verification and hypothesis to construct models. Through the corresponding relation of measurements, the SPL image was drawn and the sound field model was verified, to predict future scenarios and explore the effects of sound source energy, and then take

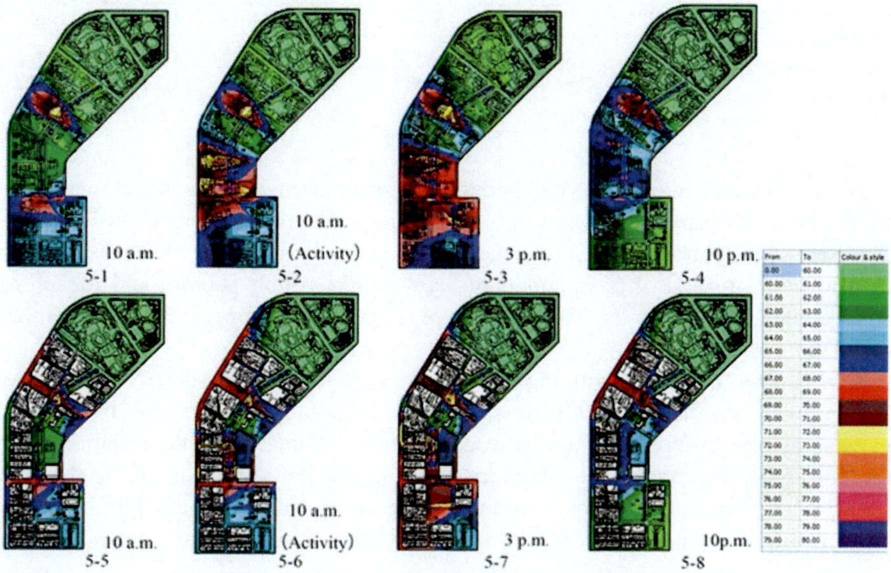

Fig. 5 Sound field simulation results of the recreational space of Caowu Trail

measures. For the sound simulation of Calligraphy Greenway from the Museum of Natural Science to Eslite Bookstore, Fig. 5-1 is the energy distribution diagram of sound field measured on a weekend morning, and Fig. 5-5 is added with surrounding building environment based on Fig. 5-1. They show that the sound field energy is mainly concentrated on Taiwan Avenue and Gongyi Road, as shown in Fig. 5-3 and 5-6. On weekends, the sound field energy of Calligraphy Greenway increases with the gathering of people, and the sound energy is mainly concentrated in eslite square and Calligraphy Greenway square. Meanwhile, as shown in Fig. 5-4 and Fig. 5-8 of computer simulation, people begin to disperse as the night falls and the sound energy decreases, which also provide a good leisure environment for residents around Calligraphy Greenway. The computer simulation results are shown in Fig. 5.

4 Evaluation and Influence of Commercial Pedestrian Streets

The questionnaire mainly consists of three parts: Part one is the background information of the respondents, including gender, age, the purpose of entering the research area. Part two is the soundscape perception evaluation scale, which is used by respondents participating in the survey to evaluate the comfort, relaxation, communication, intensity and spatial perception of the sound environment. Part three is the overall evaluation of sound perception.

Table 1 Single factor analysis table of age, frequency, and gender entering the survey area

	SS	df	MS	*F*	Sig.
Age	2.656	3	0.885	0.518	0.671
Frequency	6.059	3	2.020	5.164	0.002
Purpose	0.861	3	0.287	1.225	0.305

In the process of the survey, pedestrians on walking streets were randomly selected as the respondents, and 106 valid questionnaires were received in total. There are slightly more women than men participating in the survey, with the ratio of men to women about 4:6. Respondents are mainly young people, and the most concentrated age group is 21–30 years old, accounting for 41.5% of the total population.

Part one was evaluated with the method of single-factor analysis (ANOVA). The background information of the respondents, including age, visiting frequency, gender, and purpose of entering the research area, was analyzed in correlation with the corresponding objective indicators. The results are shown in Table 1. The significance mean is different in terms of age and gender, and age/purpose, frequency/purpose, gender/purpose, and frequency were significant (average $p < 0.05$).

4.1 Evaluation of Commercial Pedestrian Street Perception

Part two of soundscape perception was evaluated with SPSS software. Correlation analysis was made between the sound environment when the respondents participated in the survey and the corresponding objective indicators. The results are shown in Table 2. In the absence of a purpose (boring ~ interesting, uncomfortable ~ comfortable, unfriendly ~ friendly, dislike ~ like, complex ~ simple, single ~ multiple), the satisfaction of respondents is relatively high, and in (noisy ~ quiet, artificial ~ natural, chaotic ~ ordered, fierce ~ calm), respondents' perception of the current environment decreases significantly. However, after adding the purpose, shoppers and entertainers present a great contrast to players (uncomfortable ~ comfortable, unfriendly ~ friendly, dislike ~ like and artificial ~ natural, chaotic ~ orderly, fierce ~ calm), indicating that players are easy to accept the current environment.

At the same time, multivariate analysis of variance (MANOVA) was used to evaluate the second part. Correlation analysis was made between the overall evaluation of sound perception and corresponding objective indicators, and the results are shown in Table 2. There is a significance mean in (single ~ multiple) (average $p < 0.05$), and for others (noisy ~ quiet/overall evaluation, boring ~ interesting/overall evaluation, uncomfortable ~ comfortable/ overall evaluation, unfriendly ~ friendly/overall evaluation, complex ~ simple/overall evaluation, artificial – natural/overall evaluation, chaotic ~ ordered/overall evaluation, fierce ~ calm), there are differences.

Table 2 Multivariate analysis of variance for the evaluation of the soundscape perception of Caowu Trail

	SS	df	MS	*F*	Sig.
Noise~silence	12.505	3	4.168	3.500	**0.018**
Boring~fun	5.005	3	1.668	1.337	0.267
Uncomfortable~comfortable	9.467	3	3.156	2.488	0.065
Unfriendly~friendly	8.330	3	2.777	2.306	0.081
Don't like~like	1.369	3	0.456	0.333	0.801
Complexity~simplicity	10.272	3	3.424	2.223	0.090
Artificial~natural	14.401	3	4.800	2.571	0.058
A single~diverse	8.159	3	2.720	1.989	0.120
Disorder~order	12.555	3	4.185	2.671	0.051
Sharp~peace	12.103	3	4.034	2.213	0.091

SS type III sum of squares, *df* degrees of freedom, *MS* mean square, *F F* ratio, *Sig.* significance; significance (at 0.05) is in bold

5 Conclusion

Based on the theory of soundscape, this study defines the sound source type of urban soundscape and discusses the sound environment field of urban walking streets, including open-space walking space, areas along with recreation space, and commercial pedestrian streets. The research content and methods are mainly to collect and summarize the analysis of sound environment and frequency based on soundscape-related literature. The discussion of urban soundscape mainly focuses on open-space activities, and areas with fashionable and distinctive business districts, close proximity to high-density residential and with a large number of people. The characteristics of the linear green belt gather and disperse different groups of people such as commuters and loungers, thus creating linear characteristics with distinct quietness and movement.

It is very effective to use weighted SPL to represent the main factors of soundscape perception evaluation: comfort, relaxation, communication, intensity, and spatial perception. The weighted SPL can be used to manifest two of the above evaluation factors of soundscape perception, namely, communication and spatial perception.

1. Both sound source type and SPL have significant effects on soundscape evaluation. In the preference evaluation, the more polarized the sound source type is, the stronger the influence on the overall soundscape will be, and a pleasant sound source can significantly improve soundscape evaluation; meanwhile, the reduction of SPL can significantly optimize the soundscape perception of commercial streets.

2. In terms of sound source composition, man-made noisy such as talking sound and traffic sound are the dominant sound sources in the sound environment of Calligraphy Greenway recreation space at present, to which pedestrians tend to give a negative evaluation, so it should be controlled effectively; while for the initiatively designed man-made sound source such as music, pedestrians have a high degree of love, and it can be added appropriately in the state of business or activity.
3. In terms of SPL, the LAeq of Calligraphy Greenway is between 60 and 70dBA, and its equivalent SPL at 21:00 is lower than the value specified in the current specification.

References

1. S. Alves, L. Est-evez-Mauriz, F. Aletta, G.M. Echevarria-Sanchez, V.P. Romero, Towards the integration of urban sound planning in urban development processes: The study of four test sites within the SONORUS project. Noise Mapp **2**(1), 57e85 (2015)
2. T.C. Andringa, M. Weber, S.R. Payne, J.D. Krijnders, M.N. Dixon, R.V.D. Linden, et al., Positioning soundscape research and management. J. Acoust. Soc. Am. **134**(4), 2739e2747 (2013)
3. F. Asdrubali, New frontiers in environmental noise research. Noise Mapp. **1**, 1e2 (2014)
4. E. van Kempen, J. Devilee, W. Swart, I. van Kamp, Characterizing urban areas with good sound quality: development of a research protocol. Noise Health **16**(73), 380e387 (2014)
5. World Health Organization, *Guidelines for Community Noise* (WorldHealth Organization, Geneve, 1999)
6. European Parliament, & Council, *Directive 2002/49/EC Relating to the Assessment and Management of Environmental Noise* (Publications Office of the European Union, Brussels, 2002)
7. World Health Organization, *Burden of Disease from Environmental Noise* (WHO Regional Office for Europe, Copenhagen, 2011)
8. J. Kang, F. Aletta, T.T. Gjestland, et al., Ten questions on the soundscapes of the built environment [J]. Build. Environ. **108**, 284–294 (2016)
9. Alves S, Altreuther B, Scheuren J. Holistic Concept for Urban Sound Planning Applied to Real Sites[C]//Forum Acusticum, 2014
10. J. Kang, M. Zhang, Semantic differential analysis of the soundscape in urban open public space [J]. Build. Environ. **45**(1), 150–157 (2010)
11. Z. Yuan, Research on the restorative benefit of urban open space soundscape and the improvement strategy of sound environment quality [J]. New Architect **5**, 18–21 (2014)
12. Y. Yu, Z. Zhaoxi, Z. Xiaohu, et al., Street space quality measurement on human scale: A large-scale, high-precision evaluation framework combining street view data and new analysis techniques [J]. Int. Urban Plan. **34**(1), 18–27 (2019)
13. L. Jiacan, X. Lu, L. Mengyi, Z. Xie, Research on the relationship between perception and visibility of acoustic environment in pedestrian streets along main roads. Electr. Technol. **45**(5), 10–14 (2021)
14. L.-C. Chen et al., Deeplab: Semantic image segmentation with deep convolutional nets, atrous convolution, and fully connected crfs. IEEE Trans. Pattern Anal. Mach. Intell. **40**(4), 834–848 (2017)

Part II
Modern Information Technology and Management

Implementation of Machine and Deep Learning Algorithms to Detect Gaps in a Sheet Metal Industry: A Case Study

Ahm Shamsuzzoha, Timo Kankaanpaa, Huy Nguyen, and Hoang Nguyen

Abstract This research project is supported by a case company engaged in sheet metal works. The problem of the case company was to detect the minimum gaps on a sheet, as necessary to handle the sheet automatically by a gripper after punching operation. In order to solve such critical problem, the case company is provided images, which were analyzed by using machine learning and deep learning techniques. These images were consisted of the pictures of components produced from the sheet and waste materials after punching it. With human eyes, it can be observed that the gaps between some components are too small. These narrow gaps on the metal sheet after punching the predesigned components become so stuffed for the automated grippers to move the rest of the sheet safely to the warehouse or to the trash bin. Due to such small and weak gaps, automated sheer movement cannot grip properly the rest of the sheet skeleton after punching it. This results in some of the sheet parts being left behind on the worktable and needing human operators to intervene, which resulted an extra trigger and production break. In this study, an approach to machine learning and deep learning algorithms is used for the given images in order to identify which parts of the pictures are actual components and which parts are the waste material. The outcomes from this study contributes successfully to solve the gap problem of the sheet metal and find out a way to optimize the gaps between components that facilitate the easy and safe movement of the grippers and minimize the wastes materials too. This study concludes with several useful recommendations and further study perspectives in future.

Keywords Machine learning · Deep learning · Detection of gaps · Sheet metal industry · Case study

A. Shamsuzzoha (✉)
University of Vaasa, Vaasa, Finland
e-mail: ahsh@uwasa.fi

T. Kankaanpaa · H. Nguyen · H. Nguyen
Vaasa University of Applied Sciences, Vaasa, Finland
e-mail: tka@vamk.fi

C.-C. Wang, A. Nallanathan (eds.), *Proceedings of the 5th International Conference on Signal Processing and Information Communications*, Signals and Communication Technology, https://doi.org/10.1007/978-3-031-13181-3_7

1 Introduction

The goal of this research is to keep a sheet metal company's pouncing operation running smoothly. This company operates on a global basis, delivering bespoke sheet-based components. After appropriate punching operations, this organization is currently experiencing sheet material handling issues. After punching operations, waste sheets frequently stick over the worktable, failing to grip by the automated grippers completely, necessitating the use of human operators to remove the remaining sheet portions to trash bins.

It frequently causes the company's production to cease, resulting in lost output. This issue at the company arises mostly as a result of the too-narrow gaps between the components produced following the punching operation. Machine learning and deep learning methods were utilized to investigate the gaps between components with the goal of optimizing the gaps. To investigate the gaps, all of the component photos on the sheet were examined. This type of analysis of component photos aids in determining which parts of the images are actual components and which parts are waste materials to be discarded. This gap optimization approach also assists the case company in maintaining an operational schedule between the day and night production lines by allocating the most crucial and noncritical components according to the gaps.

The gaps between components are investigated in this study by converting the component images from vector to pixel images using a machine learning technique. This image conversion technique aids in the study and helps the organization analyze component design and gaps for a more stable and efficient manufacturing process. The following is how the rest of the article is structured: Section 2 discusses some recent machine learning research, whereas Sect. 3 discusses the research methods. Section 4 contains a description of the case company, and Sect. 5 contains the study results. Section 6 concludes this research study by acknowledging the study's contributions.

2 Literature Review

Because of digitalization of manufacturing industry, it is essential nowadays to introduce digital tools and technologies in this segment. This digitalization process results in enormous amount of valuable data which needs to be analyzed in order to improve production processes and detecting causes of problems. In this environment, machine learning and deep learning can be useful to save energy, time, and resources and avoid waste [1, 2].

The term machine learning describes algorithms used to identify and extract valuable data pattern to improve quality. Implementation of machine learning is not a new field. However, recent growth in computational power, many industries have been considering employing artificial intelligence solutions to improve production processes [2, 3]. The development of machine learning has been incredibly successful at tackling complex tasks [4].

Machine learning and deep learning algorithms are successfully applied to sheet metal industries for selecting a suitable manufacturing process to achieve the final geometry of a metal part that is unstructured and heavily reliant on human expertise [5, 6]. Machine learning is also used to predict defects in sheet metal forming processes [7, 8].

3 Study Methodology

The aim of this study was to discover the appropriate spacing between components in order to easily manage the waste sheet following a punching operation in the sheet metal sector. Machine learning and deep learning algorithms were used to evaluate the photos of sheets provided by the sheet metal industry in order to identify and show the gaps. Machine and deep learning methods were used to turn the provided photos into suitable ones that could be easily evaluated.

The machine learning system was given an image of a metal sheet with components on it and a threshold, which is the distance between any two components that is acceptable. Furthermore, the provided image is a grayscale.svg file. The algorithm's output is the same image as the input, with a circle indicating the location of a too narrow gap (less than the chosen threshold) between components. The final image should include every feasible narrow area. The relevant phases of the machine learning method are programmed in Python 3.7.0 in this study. Several libraries were used during this project.

3.1 Machine Learning Algorithm Description

- Step 1: Convert the input image from a.svg (vector) file to a.jpg or.png file (pixels image).
- Step 2: Obtain component coordinates from the pixel image input.
  Figure 1 shows the program required to convert a vector image to a pixel image. The transformed image is shown in Fig. 2.
- Step 3: Dispose the component's body.
  The main goal of this technique is to figure out how far apart components are. The distance between the boundaries of those components, in particular, must be measured. As a result, there is no need to look after your physique. Furthermore, removing the body aids in reducing the algorithm's running time. Figure 3 shows the program required to determine the component's border.
- Step 4: Make the detection.

```
def _Get_components(self, show_components = True):
    """Process binary image to get each single
    machine coponents obj
    Parameters:

    ------------
    Returns:
    ------------
    num_component: int, numbers of component
    components: list, contains obj components

    """
    # cv2.namedWindow("output", cv2.WINDOW_NORMAL)
    components=[]
    # Perform the operation
    ret, labels, stats, centroids = cv2.connectedComponentsWithStats(self.image_binary)
    # ret already include background, we should minus 1 to get
    # num ber of component
    num_component = ret - 1

    # Loop through all label
    for label in range(2, ret):
        # Creat mask of each single component
        mask = (labels == label).astype(np.int32)
        # Get index of mask
        ids = np.where(mask == 1)
        # Get boundary
        boundary = self._boundary(ids)
        components.append(Component(ids, centroids[label], boundary, self.img_size))
    return components, num_component
```

Fig. 1 Function to get the coordinates of components

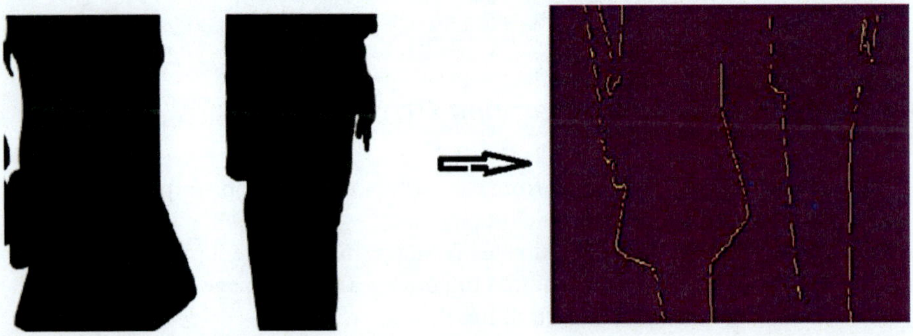

Fig. 2 Transformation of vector image to pixel image

Assume that the lowest permissible distance (threshold) is d (pixels). We draw a circle with a radius of $d/2$ on every pixel p in the boundary. As a result, if any two circles from distinct components overlap, this is the spot where we need to make a note (because at this location, the distance of two components is less than d). Figure 4 depicts several circles over each component as needed to locate gaps between them.

```
def _boundary(self, coordiante):
    """Process binary image to get each single
    machine coponents obj
    Parameters:
    coordinate: tuple, the first element is
    numpy array of y axis and second element is
    also numpy array of x axis
    ------------
    Returns:
    ------------
    ids: tuple, the first element is numpy array
    of y axis and second element is also numpy array
    of x axis. These coordinate represent boundary of
    each component

    """
    # Create temp to show component
    temp = np.zeros(self.img_size, dtype=np.int32)
    temp[coordiante] = 1
    # Perform operation
    temp = binary_dilation(temp == 0, self._k) & temp
    # get ids
    ids = np.where(temp == 1)
    return ids
```

Fig. 3 Function to get the boundary of components

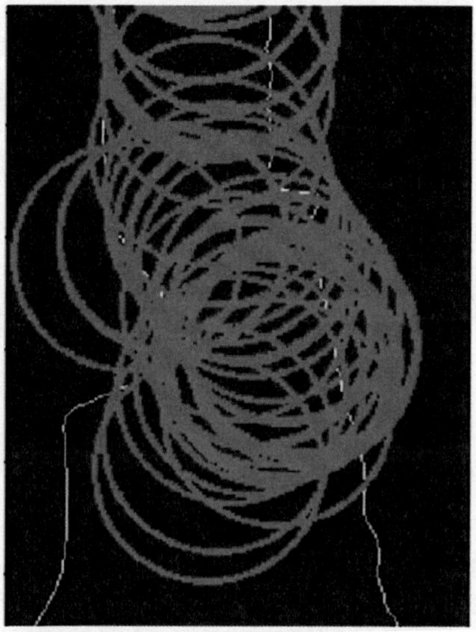

Fig. 4 Various boundary circles over a single component

3.2 Complexity

Let's assume after converting input from vector image to pixels image, the input becomes an image with the resolution $m \times n$. The algorithm basically goes to every single pixel in input to get the boundary and plot circle; therefore, the algorithm has O ($m \times n$) running time.

4 Description of the Case Company

The case company is a leading specialist in machines and systems for sheet metal working. This company offers one of the widest and covers all applications: laser processing, punching, shearing, bending, and automation. The company has numerous locations in Italy, Finland, the United States, and China, from which it ships machinery and systems all over the world. The sales and service network of the company operates in over 80 countries, either directly or through a network of specialist dealers.

The product lines in the company are ThePUNCH, TheLASER, TheCOMBI, TheBEND, TheSYSTEM, and TheSOFTWARE, which cover all stages of the sheet metal working process. This product family also offers highly advanced servo-electric solutions for punching, bending, and integrated processes that is the widest in the world. All the products from the company are developed according to the "green means" concept, combining sustainability and productivity.

5 Result Analysis

The result image is as shown in Fig. 6 after applying the input image as shown in Fig. 5 with the parameter threshold 1.5 pixels. The output shows that the algorithm performs admirably. It precisely recognizes spots with a width of less than 1.5 pixels (the circles in Fig. 6). Furthermore, it returns the exact border and removes the component's body.

5.1 Sensitivity to Noise

Because the suggested approach examines each and every pixel in the input, it is extremely sensitive to noise, which could be variations in brightness or color information in the image input. Preprocessing the input image helps fix this problem.

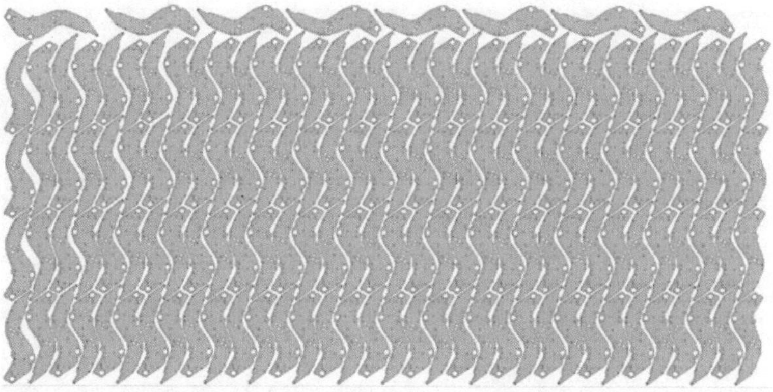

Fig. 5 Input image

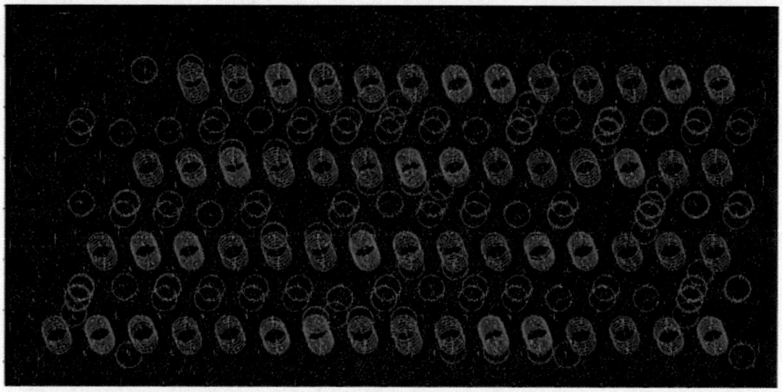

Fig. 6 Output image

5.2 Running Time

As previously stated, this algorithm takes O ($m \times n$). In general, this is an excellent application. This technique takes a little longer to run when the input is more sophisticated (more noise, more components). This difficulty can be handled by vectorizing the algorithm's operations. By omitting the loop and instead employing vectorized arithmetic, as demonstrated in Fig. 7, you can have a faster running time.

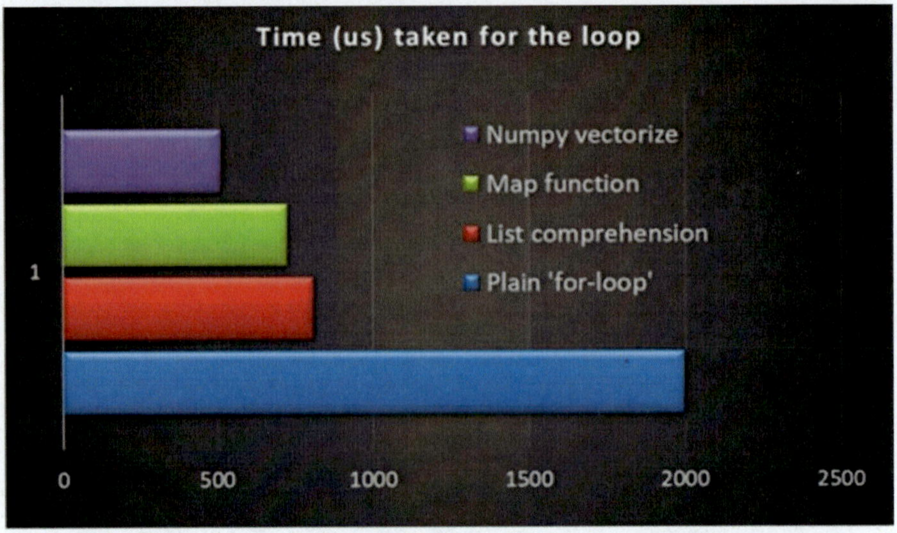

Fig. 7 Comparative speeds of execution for conditional loop-based code block

5.3 *Analysis of the Gaps Between the Components*

In order to punch the components after punching on the metal sheet, the gaps between each component are critical. When the spaces between components are too small, it causes a problem on the work table by leaving part of the sheet pieces on it. These remaining portions stick to the worktable and cause the automated grippers to cease moving the waste sheet to the garbage bin. This also causes the entire punching operation to shut down, resulting in a loss of productivity. The case company's goal was to reduce the spaces between components so that waste metal sheet would not stick to the worktable.

In order to optimize the gaps between punched components, some pictures attached with the design of the components over the sheet metal were taken into consideration to analyze the gaps. These images were in vector format, and the machine learning method translated them to pixel images, as shown in Fig. 8. This conversion increases the visibility of the images in order to minimize the gaps between the components.

Figure 9 depicts the 1.5-pixel space between two neighboring components, which satisfies the design criteria and leaves no room for error, as seen in Fig. 10.

Figure 11, on the other hand, shows component gaps of more than 1.5 pixels, which is an unacceptable design requirement that causes a difficulty on the worktable at the example company. Several faults are discovered following analysis and visualized in Fig. 12 as a result of such conflicts with the design criteria.

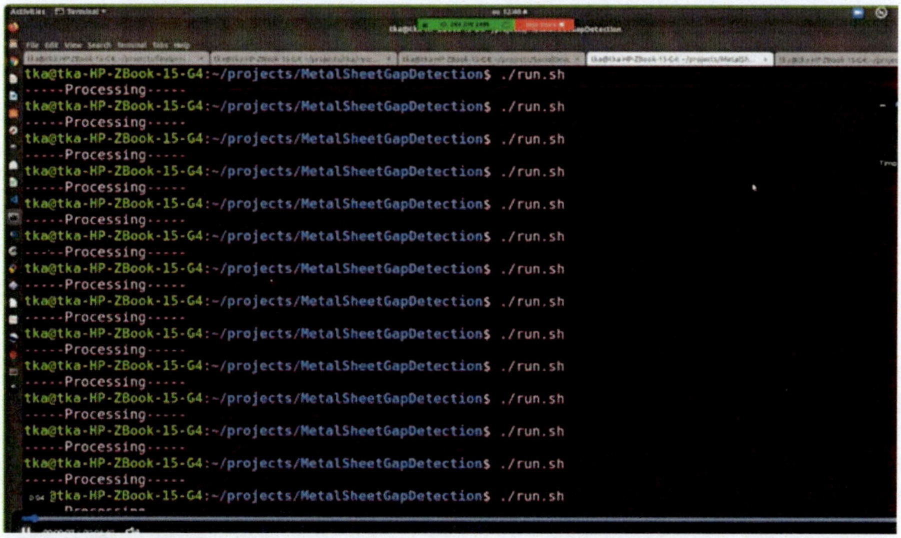

Fig. 8 Sample program to analyze the gaps between components

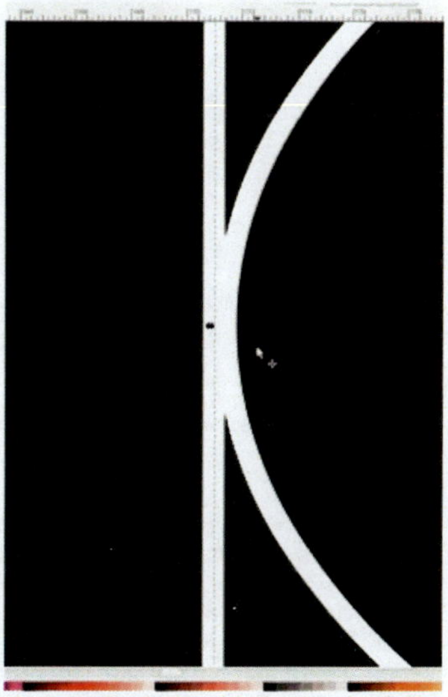

Fig. 9 Visualization of tolerable gaps between components

As long as optimal spacing between components are not achieved, such faults will continue to rise. As previously stated, this optimization is accomplished through the use of machine and deep learning algorithms. It essentially makes use of computer vision and image processing techniques to help eliminate gaps utilizing the approach described in Sect. 3. After appropriate processes, this optimization method facilitates waste sheet management. It also helps to make necessary shifting of punching operations on sheet metals between night shift when the operation is

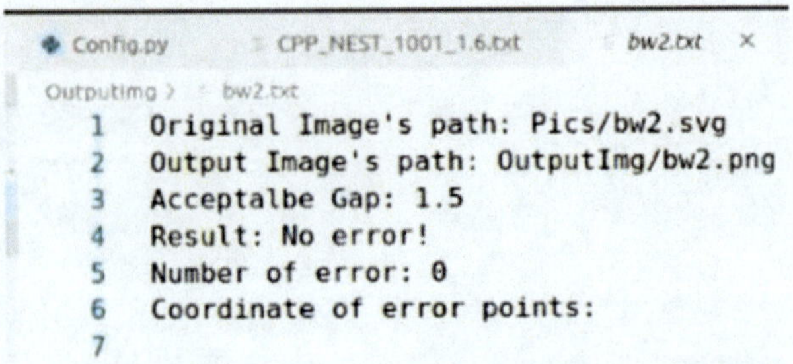

Fig. 10 Visualization of number of errors after assigning specific gaps between components

Fig. 11 Visualization of non-tolerable gaps between components

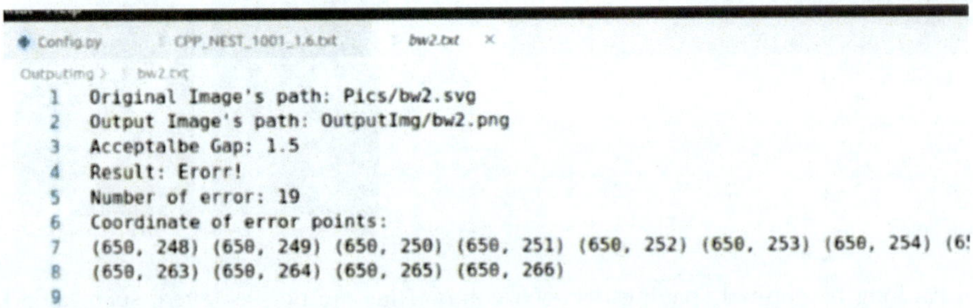

Fig. 12 Visualization of the number of errors after assigning specific gaps between components

done fully automated and unmanned and day shift when the operation is supported partially by human interventions. This shifting decision promotes the company to save its costly resources as well as enhance productivity in general.

6 Discussion and Conclusions

Component analysis was performed to develop and limit gap detection between components from photos during the investigation. The proposed algorithm appears to function well in this investigation, with the exception that some issues, such as noise sensitivity and speeding up the running time, still need to be addressed in order to achieve better results. This research analyzes and contributes to the operational restrictions of a Finnish case sheet metal industry.

The use of a component analysis technique developed through machine learning improves the company's operational flexibility and reduces production time. It ensures that the company's shifts are maintained throughout the day and night. For example, complex components with important gaps between components were isolated and reserved for the day shift, when human operators can assist in removing the waste sheet after the punching operation if the automated gripper is blocked. The night shift, on the other hand, is dedicated to less intricate components with appropriate intervals between them, where automated grippers comfortably remove the waste sheets after operation. As a result, this gap optimization technique, which uses machine learning, helps to maintain production stability and quality at the same time.

Acknowledgments Authors like to acknowledge the active support from the case company to conduct this research study. Essential financial support is also highly appreciated.

References

1. S.-Y. Tsai, J.-Y. Chang, Parametric study and design of deep learning on leveling system for smart manufacturing, in *IEEE International Conference on Smart Manufacturing, Industrial & Logistics Engineering (SMILE)*, (Hsinchu, 2018)
2. D. Weichert, P. Link, A. Stoll, S. Ruping, S. Ihlenfeldt, S. Wrobel, A review of machine learning for the optimization of production processes. Int. J. Adv. Manuf. Technol. **104**, 1889–1902 (2019)
3. T. Ademujimi, M. Brundage, V. Prabhu, A review of current machine learning techniques used in manufacturing diagnosis, in *IFIP International Conference on Advances in Production Management Systems*, (2017), pp. 407–415
4. M. Sharp, T. Hedberg, A survey of the advancing use and development of machine learning in smart manufacturing. J. Manufact. Syst. **48**, 170–179 (2018)
5. E. Hamouche, E. Loukaides, Classification and selection of sheet forming processes with machine learning. Int. J. Comput. Integr. Manuf. **31**(9), 921–932 (2018)
6. M.-C. Chiu, C.-D. Tsai, T.-L. Li, An integrative machine learning method to improve fault detection and productivity performance in a cyber-physical system. J. Comp. Inform. Sci. Eng. **20**(2), 12 (2020)
7. M. Deb, B. Ribeiro, P. Prates, Model prediction of defects in sheet metal forming processes, in *International Conference on Engineering Applications of Neural Networks*, (2018), pp. 169–180
8. S.-Y. Tsai, J.-Y. Chang, Design of deep learning on intelligent levelling system for industry 4.0 technology, in *MATEC Web of Conferences, 185, 00026*, (2018)

Modelling the Service Process at the Cash Register Using Generating Functions

Andreas Ahrens, Detlef Hartleb, and Jelena Zăšcerinska

Abstract Queue management is an important part in our life as it helps smooth out daily activities. Typical cases for use of queue management are the avoidance of queues at the cash register in grocery stores or the reduction of traffic jams on the roads. Often customer arrival in the queue is bursty. Therefore, such a queue has to be optimized. In this work, a new model for the analysis of buyers' service based on gap processes is presented, and the service process is described by generating functions. As a practical application, the check-out process at the cash register consisting of the buyer waiting in the queue to the cash register and the buyer waiting for the service needed for scanning the goods and payment (also referred to as payment processing) is studied. The obtained closed form solutions allow concluding on how the level of burstiness in the two individual processes influences the buyers' service at the cash register. For a practical verification, the data of two grocery shops in Lithuania are studied. The obtained results show that burstiness is a limiting factor when analyzing the service effectiveness at the check-out. Furthermore, the bursty nature of customers in the queue to the cash register prevails in the service process and finally influences the service quality significantly.

Keywords Customers' burstiness · Cash register · Customers' waiting time · Gap distribution function · Generating functions

A. Ahrens (✉)
Hochschule Wismar, University of Technology, Business and Design, Wismar, Germany
e-mail: andreas.ahrens@hs-wismar.de; https://www.hs-wismar.de

D. Hartleb
Hochschule Wismar, University of Technology, Business and Design, Wismar, Germany

ETSIST, Universidad Politécnica de Madrid, Madrid, Spain
https://www.etsist.upm.es/

J. Zăšcerinska
Hochschule Wismar, University of Technology, Business and Design, Wismar, Germany

Centre for Education and Innovation Research, Riga, Latvia

C.-C. Wang, A. Nallanathan (eds.), *Proceedings of the 5th International Conference on Signal Processing and Information Communications*, Signals and Communication Technology, https://doi.org/10.1007/978-3-031-13181-3_8

1 Introduction

Queues can emerge in many areas of our daily life such as in banking, human resource management, road traffic and telecommunication systems [1–3]. Most of them operate according to a simple first-in first-out (FIFO) principle, i. e. customers who arrive first are served first. Examples of this type of queueing can be found, e. g., at the airport check-in or in the grocery store. Although operators have more information than customers – e. g., airlines know when each upcoming flight will leave or shop-operators know rush-hour times – they do not take an advantage of it. In such situations, a bursty arrival of customers can lead to a poor queue performance, e. g. when checking in at the airport or checking out at the grocery shop. The concentrated arrival of customers will lead to a highly fluctuating arrival rate over the time. As a consequence, long waiting times appear since many people enter the queue at the same time.

Many queueing systems have been proposed and analysed in the literature [4, 5]. While the literature on queueing theory is very large, the use of operators' knowledge to decrease waiting times appears to be under-explored. Queuing systems went from simple physical barriers to the state-of-the-art digital applications. In [6], a signalling system for decreasing customers' waiting based on the operators' knowledge is investigated and [7] focuses on client server management in grocery shops.

The key elements of such kind of queuing systems are customer and server. A customer refers to anything and anyone that arrives at a facility and requires service. Examples of customers are shoppers at the check-out station or cars at the traffic light. In turn, a server means a resource that provides the requested service, e. g. a technician or a check-out station at a grocery shop. Table 1 illustrates the relationship between systems, customers and servers.

Regarding the term 'customer', a variety of definitions exists. It should be pointed that the terms 'customer' and "client' are used synonymously in the scientific literature. The term 'lost customers' is introduced by Sharma [8], and also customer abandonment behaviour [9] is used with regard to a customer who stopped trying to receive a service due to a number of reasons. The term 'blocked customers' describes one who retries to get a service after a random amount of time [8]. Hence, the researchers have established the understanding of the customer behaviour based on the reception of service. Against this background, the authors of the present work base their research on the concept of binary customer behaviour, namely, to buy or not to buy [10]. In the present work, a customer is the overall term meaning anyone who arrives at a shop and requires a service. Customers are differentiated into visitors (i.e. who visit a shop but do not buy anything) and buyers (who visit a

Table 1 Examples of systems, customers and servers

System	Customer	Server
Grocery shop	Shoppers	Check-out station
Hospital	Patients	Nurses
Road network	Cars	Traffic light
Reception desk	Guest	Receptionist

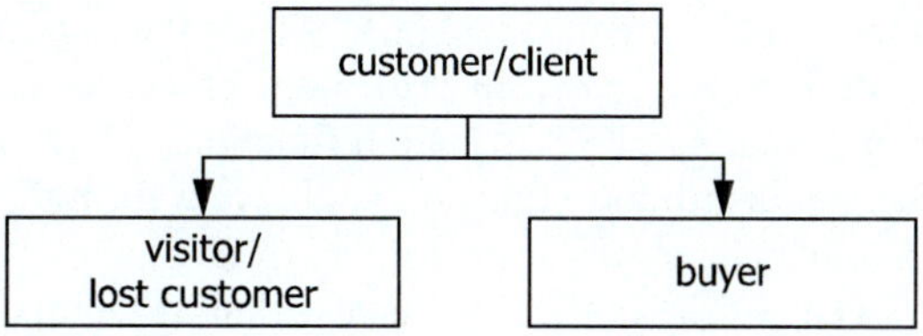

Fig. 1 The relationship between customer/client, visitor and buyer

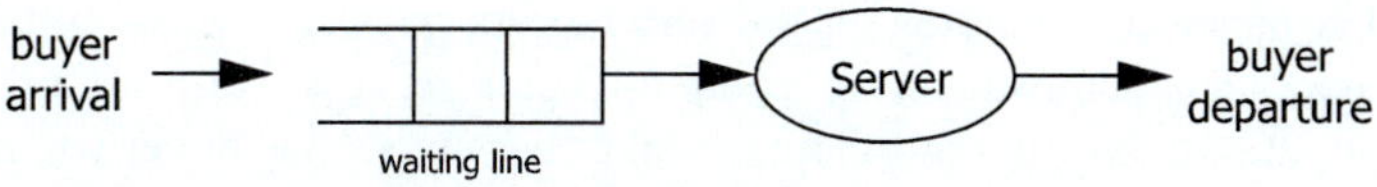

Fig. 2 Phases of the service process at the cash register

shop and buy a product) [10]. Buyers are further differentiated into two types: (1) highly delay-sensitive but only check-out a few items and (2) highly delay-insensitive but check-out many items [11]. Consequently, the overall term 'customer' is used in the same meaning by Sharma [8] and the authors of the present work, while the terms 'lost customer' and 'visitor' are used synonymously. Figure 1 illustrates the relationship between the definitions of customer/client, visitor and buyer.

Another key element in the queuing system is a server. By a server, a station that provides service is understood [8]. The server deals with units and items. Service being delivered to a customer is a process. The service process can be divided into phases [12]. Such modelling enables the search for an appropriate balance between the cost of service and the duration of waiting [13].

The literature analysis leads to the understanding that queuing systems are based on three key phases. The three key phases proceed from customer arrival phase (phase 1) [6, 8] through the service phase (phase 2) to the departure phase (phase 3) [8].

The previous research in the field of the queueing model carried out by Choudhury and Deka [14] focused on the joint analysis of two phases of service, namely Bernoulli vacation and server breakdown, during the service. However, this work identifies the service process, namely the payment process at the cash register, as the unity of two processes: (1) waiting in the queue to the cash register and (2) payment processing at the cash register [10]. Figure 2 illustrates the service process at the cash register considered in this paper, whereas in [2], the payment processing was divided into two phases: (1) scan and pay and (2) payment.

Against this background, this paper shifts the research from the check-in/arrival [6] and queue management [7] to the service process as the unity of waiting in the queue to the cash register and the payment processing at the cash register. In comparison to [2], the focus is put on the concentrated buyers' arrival in the queue to the cash register as the analysis of the parameters – waiting time and payment time – presented in [2] does not give any information on the buyers' burstiness in the service process. As a practical example, the check-out process consisting of the arrival in the queue to the cash register as well as the service needed for scanning the goods and for the payment (also referred to as payment processing) at a grocery shop is studied. As shown in [10], both processes can be described by gaps and their corresponding gap distribution functions. Here, a bursty behaviour could emerge if the gap distribution function is different from the exponential distribution function. Possible characteristics are given by Wilhelm or Weibull for bursty processes [15].

The novelty of the paper is revealed by defining the service process as the unity of waiting in the queue to the cash register and the payment processing at the cash register. This paper concentrates on the service (check-out) process by taking the individual gap's characteristics into account. Therefore, it is worth noting that this paper investigates on how the individual level of burstiness (i.e. buyer waiting in the queue to the cash register and buyer paying at the cash register) influences the service process as a whole. This approach will allow modelling and optimizing the service process based on the unity of waiting in the queue and payment processing. For the mathematical formulation, the concept of generating functions is used [10].

The remaining part of this paper is organized as follows: bursty business processes and the corresponding modelling by gaps are introduced in Sect. 2. The serving of the buyers at the check-out is analysed in Sect. 3 followed by a discrete model of the buyer's service processing presented in Sect. 4. The obtained simulation results are evaluated in Sect. 5. Finally some concluding remarks are given.

2 Bursty Business Processes and Gap Functions

When analysing the spread of buyers within a sequence of non-buying visitors (a visitor becomes with a probability p_e a buyer), the distribution of the buyers is of high interest for providing each buyer a good service by, e.g., avoiding queues (see also Fig. 3). The distribution of the buyers can be described by gaps, and the corresponding gap distribution function can be used to analyse the visitor–buyer

Fig. 3 Buyers' burstiness (represented by "x") within a sequence of shop visitor (represented by"-")

```
xx-xxx--xxxxxx-xx------------
-------xxxx-x--xx--xx--------
-----------------------------
-----------------------------
-----------------------------
-----------------------------
-----------------------------
-------------------------xxxx
-----------------------------
-----------------------------
-----------------------------
-----------------xx----------
-----------------------------
--------------xxxx-----x--xx-
-----------------------------
-----------------------------
----xxx----xxx---------------
-----------------------------
-------xxx---------------xx--
-----------------------------
xxxxx-----------xx--xxxxxx-x-
-----------------------------
---xxxxxx--------------------
```

relationship. Practical investigations have shown that Weibull or Wilhelm-based gap distribution functions achieve good results especially in bursty situations [15]. Bursty as well as non-bursty processes can be described by a gap distribution function $u(k) = P(Y \geq k)$ defining the probability of a gap Y greater than or at least equal to a given number k or a gap density function $v(k) = P(Y = k)$ representing the probability of a gap Y equal to a given number k. An alternative representation to $u(k)$ is given by the generating function $U(t)$ representing the coefficients of $u(k)$ by a power series [16, 17].

An appropriate choice for $u(k)$ is given by Wilhelm and defined as

$$u(k) = [(k+1)^{\alpha} - k^{\alpha}] \cdot \mathrm{e}^{-\beta \cdot k}. \tag{1}$$

The function $u(k)$ defined in (1) depends in analogy to Fig. 3 on the buyer probability p_e and the buyer concentration $(1 - \alpha)$. The buyer probability can be defined as

$$p_e = \frac{E\{\text{number of buyers}\}}{E\{\text{number of visitors}\}} \tag{2}$$

with the parameter $E\{\cdot\}$ denoting the expectation functional. The parameter β in (1) is defined as $p_e \approx \beta^{\alpha}$ [15, 18]. Practically, relevant buyer concentration is in the range of $0.0 < (1 - \alpha) \leq 0.5$, whereas the buyer concentration of $(1 - \alpha) = 0$ describes the situation with non-bursty buyers (also refers to memoryless buyer scenario), where the buyer probability p_e is sufficient to describe the buying process. The generating function belonging to this gap-distribution function results according to [10] in

$$U(t) = \frac{1}{(1 - \mathrm{e}^{-\beta} t)\alpha}. \tag{3}$$

The generating function, defined in (3), can be used to calculate the elements of the corresponding gap distribution function $u(k)$ in analogy to the gap distribution function defined in (1). Given the generating function

$$U(t) = \sum_{k=0}^{\infty} u(k) t^k \tag{4}$$

the corresponding elements of the sequence $u(k)$ can be calculated based on Taylor's theorem by the k-th derivative of the generating function $U(t)$ at $t = 0$. The elements of the sequence $u(k)$ result in

$$u(k) = \frac{\Gamma(\alpha + k)}{\Gamma(1 + k) \cdot \Gamma(\alpha)} \mathrm{e}^{-\beta k} \tag{5}$$

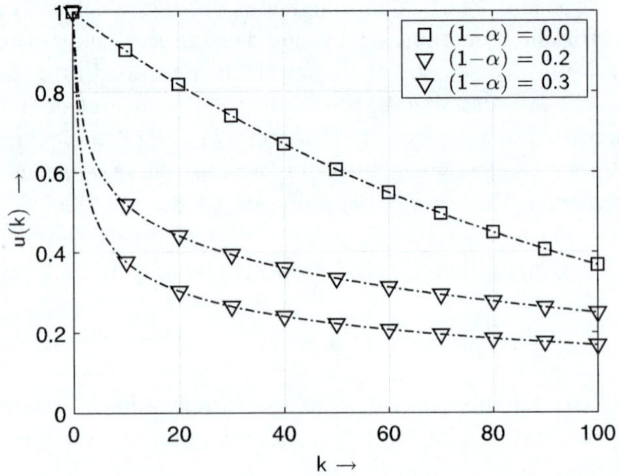

Fig. 4 Gap distribution function $u(k)$ according to (5) assuming a buyer probability of $p_e = 10^{-2}$

as shown in [10]. Figure 4 highlights the characteristic of the gap distribution function $u(k)$ for different parameters of $(1 - \alpha)$. The gap distribution function $u(k)$ defined in (5) shows a similar behaviour compared to the gap distribution function $u(k)$ defined in (1). The bursty behaviour is expressed by the non- exponential characteristic obtained for values of $(1 - \alpha) > 0$.

However, for a good customer service, not only the visitor–buyer relationship is of any interest but also the avoidance of queues or delays during the service process, e.g. when checking out. When analysing the service process at the check- out, it has been shown that the service process can be divided into the process of waiting in the queue to the cash register and the payment processing (e.g. scanning of the goods and the payment) at the cash register. Both processes have been analysed and modelled based on the data obtained from two supermarkets in Lithuania in [10, 12] with the result that the waiting times in the queue to the cash register have a more burst-like structure compared to the buyers' payment processing time. Solutions for estimating the expected level of burstiness are presented in [10, 19].

Unfortunately, the independent modelling of the two processes does not give any results for increasing customer satisfaction. Therefore, this paper looks at the service process as the whole (i.e. waiting in the queue to the cash register and in the payment processing) and how the individual level of burstiness in the two gap processes influences the service process. The concept of generating functions allows merging of the two independent gap processes (i.e. waiting time in the queue to the cash register and the payment processing time at the cash register) when calculating the probability that a buyer is served within a given time interval at the check-out.

3 Buyer's Service Processing at the Check-Out

The buyer's check-out is divided into two parts: waiting in the queue to the cash register and the payment processing at the cash register (see also Fig. 5). Each of the processes can be modelled by gaps.

The waiting in the queue to the cash register, defined by the gap distribution function $u_{wt}(k)$, can be described by the probability p_{wt} defined by the average length of free time intervals between two buyers at the cash register ${m_1}^{(wt)}$ and the corresponding level of burstiness $(1-\alpha_{wt})$.

The average gap length or average length of free time intervals between two buyers in the waiting line to the cash register defined by p_{wt} can be expressed as

$$\frac{1}{p_{wt}} - 1 = m_1^{(wt)} = \sum_{k=0}^{\infty} k \cdot v_{wt}(k). \tag{6}$$

Accordingly, the payment processing time at the checkout, defined by the gap distribution function $u_{pt}(k)$, can be described by the probability p_{pt} identified by the average time of the buyers' payment processing at the cash register ${m_1}^{(pt)}$ and the corresponding level of burstiness $(1-\alpha_{pt})$. The average payment processing time at the cash register defined by p_{pt} can be expressed as

$$\frac{1}{p_{pt}} - 1 = m_1^{(pt)} = \sum_{k=0}^{\infty} k \cdot v_{pt}(k). \tag{7}$$

It is worth noting that the probabilities p_{wt} and p_{pt} are different from the buyer probability p_e defined in (2) and used in (1) as the process of service is divided into two parts. The same is valid for the buyers' concentration.

Assuming that the two gap processes X and Y are independent with the probability distributions $P(X \geq j) = u_{wt}(j)$ and $P(Y \geq k) = u_{pt}(k)$. The event $(X \geq j, Y \geq k)$ has the probability $u_{wt}(j) \cdot u_{pt}(k)$. The sum $S = X + Y$ is a new random variable, and the event $S = r$ is the set of mutually exclusive events

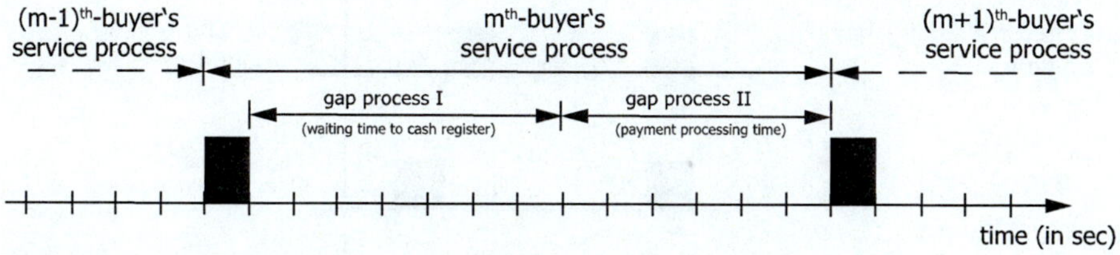

Fig. 5 Modelling the payment process by two independent gap processes

$$(X \geq 0, Y \geq r), (X \geq 1, Y \geq r-1), \ldots, (X \geq r, Y \geq 0) \tag{8}$$

The distribution $u(r) = P\,(S = r)$ results in

$$u(r) = u_1(0) \cdot u_2(r) + \ldots + u_1(r) \cdot u_2(0). \tag{9}$$

As X and Y are independent variables with the generating function $U_{\mathrm{wt}}(t)$ and $U_{\mathrm{pt}}(t)$, the sum $S = X + Y$ is defined by the generating function $U\,(t) = U_{\mathrm{wt}}(t) \cdot U_{\mathrm{pt}}(t)$. The corresponding gap distribution function can be obtained by the convolution of the corresponding gap distribution functions $u_{\mathrm{wt}}(k) * u_{\mathrm{pt}}(k)$.

4 Discrete Model of the Buyer's Service Processing

Figure 6 illustrates the corresponding system for modelling the buyers' service at the check-out. The service process of an individual buyer depends on the waiting time in the queue to the cash register defined by the gap distribution function $u_{\mathrm{wt}}(k)$ and the payment processing time at the cash register (also referred to as the buyers' payment processing time) defined by the gap distribution function $u_{\mathrm{pt}}(k)$.

The probability of processing a single buyer within an interval of the length $n = 3$ is illustrated in Fig. 7. The buyers' service is defined from a short waiting in the queue to the cash register and a long buyers' payment processing time (defined by

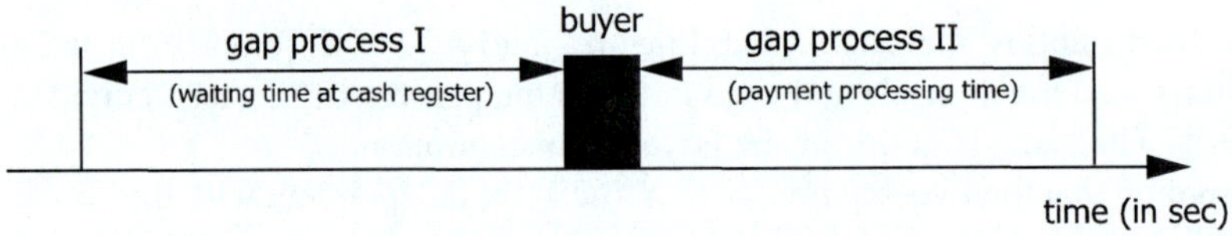

Fig. 6 Analysing the probability of a single buyer processed in an interval of the length $n = a + b + 1$ (with a = gap length process I and b = gap length process II)

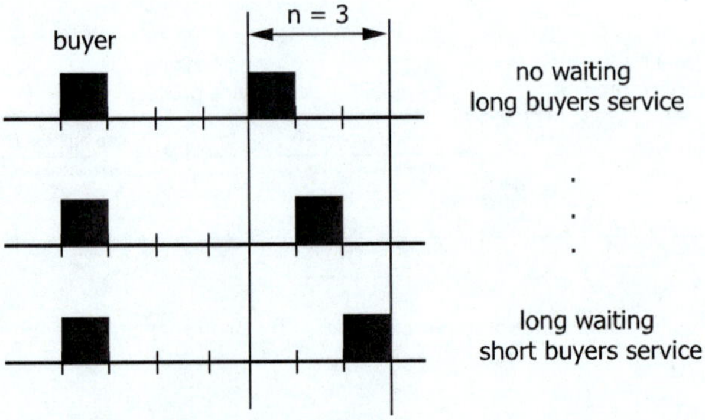

Fig. 7 Analysis of the probability of a single buyer processed within an interval of length $n = 3$

the probability $p_e \cdot u_{wt}(0) \cdot u_{pt}(2)$) to a long waiting time in the queue to the cash register and a minimum buyers' payment processing time (defined by the probability $p_e \cdot u_{wt}(2) \cdot u_{pt}(0)$).

Therefore, the probability of having a single buyer served within a given interval n is given by

$$P(n, g = 1) = p_e \cdot \left(u_{wt}(0) \cdot u_{pt}(n-1) + \ldots\ldots + u_{wt}(n-1) \cdot u_{pt}(0)\right). \quad (10)$$

The convolution of the gap distribution functions $u_{wt}(k)$ and $u_{pt}(k)$ contained in (10), i. e. $u_{wt}(k) * u_{pt}(k)$, can be described by the product of the corresponding generating functions and results in

$$U(t) = U_{wt}(t) \cdot U_{pt}(t). \quad (11)$$

As shown in [10], each of the gap distribution functions can be described by a generating function with different parameters.Assuming that the two gap processes have identical parameters such as

$$p_{wt} = p_{pt} = p_e \quad (12)$$

and

$$\alpha_{wt} = \alpha_{pt} = \alpha \quad (13)$$

the generating function $U(t)$, defined in (3), results in

$$U(t) = \frac{1}{(1 - e^{-\beta}t)^{2\alpha}}, \quad (14)$$

with the parameter concentration $(1 - \alpha)$ and $p_e = \beta^{\alpha}$. According to [10], the gap distribution function belonging to this generating function can be obtained, and, therefore, the probability of processing a single buyer within a time interval n (for $n = 1, 2, 3, \ldots$) results in

$$P(n, g = 1) = p_e \cdot \frac{\Gamma(2\alpha + n - 1)}{\Gamma(n) \cdot \Gamma(2\alpha)} e^{-\beta(n-1)}. \quad (15)$$

For $n = 1$, the probability $P(n = 1, g = 1)$ results in

$$P(n = 1, g = 1) = p_e \quad (16)$$

and equals, as expected, the buyer probability p_e. Furthermore, with $\alpha = 1$ (memoryless buyer scenario) the probability $P(n, g = 1)$ results with

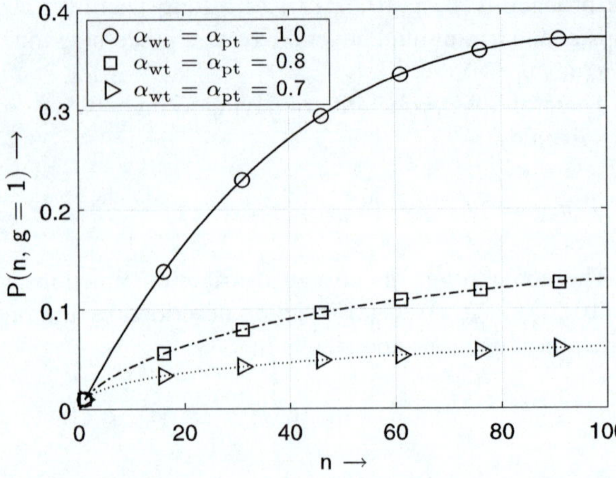

Fig. 8 Analysis of the probability of a single buyer processed assuming identical levels of burstiness at probability of $p_{wt} = p_{pt} = 10^{-2}$

$$\frac{\Gamma(n+1)}{\Gamma(n)} = n \tag{17}$$

and $\beta = p_e$ and $1 - p_e \approx e^{-pe}$ in

$$P(n, g = 1) \approx n \cdot p_e \cdot (1 - p_e)^{n-1} \tag{18}$$

and coincides with the Bernoulli model. Figure 8 highlights the obtained results when both gap processes have identical levels of burstiness and mean values. It shows that the probability of serving a single buyer within a given time frame n decreases with the increasing level of burstiness. Therefore, it can already be concluded that burstiness is a limiting factor when analysing the buyers' service.

5 Results

The service process can be divided into two parts: waiting in the queue to the cash register and processing of the bought goods (the so-called payment processing time). The question of optimising the check-out leads to the question of how the level of bustiness in the two processes as well as the mean value of the gaps between the buyers when waiting in the queue to the cash register and the average payment processing time influences the flow of buyers.

By analysing the data of two supermarkets in Lithuania, the parameters of the two shops are found to be quite different. In [10, 12], the following limits were found:

$$m_1^{(wt)} > m_1^{(pt)} \text{ or } p_{wt} < p_{pt} \tag{19}$$

and

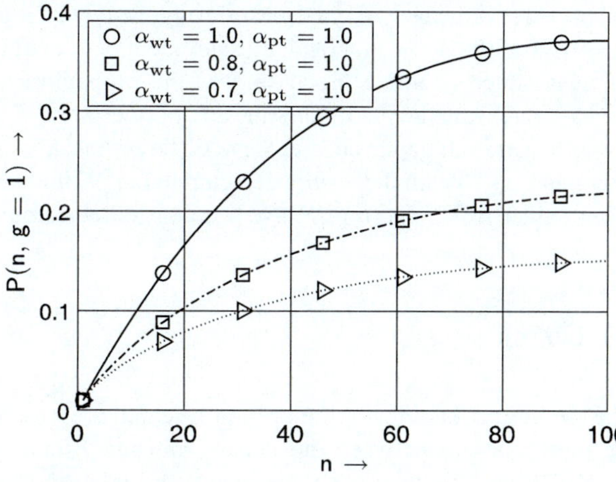

Fig. 9 Analysis of the probability of a single buyer processed in a given time interval *n* assuming different levels of burstiness at a probability of $p_{wt} = p_{pt} = 10^{-2}$

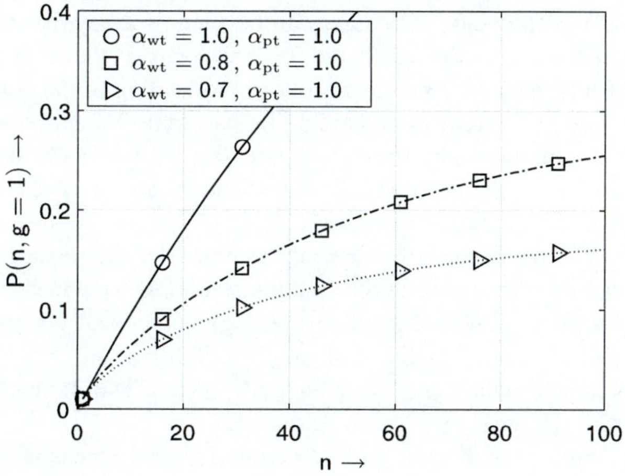

Fig. 10 Analysis of the probability of a single buyer processed in a given time interval *n* assuming different levels of burstiness at a probability of $p_{wt} = 10^{-3}$ and $p_{pt} = 10^{-2}$

$$\alpha^{(wt)} \approx 0.7 \ldots 0.9 \text{ and } \alpha^{(pt)} \approx 1. \tag{20}$$

Here, it is worth noting that the analysis of the payment processing time has not shown any significant level of burstiness, i. e. $\alpha^{(pt)} \approx 1$ when analysing the data of the two supermarkets in Lithuania [12]. Figures 9 and 10 show the obtained results by analysing the probability of serving a single buyer in a given interval *n*, i.e. $P\,(n,\, g = 1)$. Figure 9 highlights the obtained results with different levels of burstiness and identical mean values. Here, it is assumed that the buyers' service (i.e. scanning of the goods and the payment) does not create any burstiness as confirmed by practical measurements. Compared with the case shown in Fig. 8, where both processes had a certain bursty nature, it can be confirmed that the probability $P\,(n,\, g = 1)$ decreases with an increasing level of burstiness $(1 - \alpha_{wt})$, i.e. $\alpha_{wt} < 1$. From this, it can be concluded that the lack of burstiness during the payment process (i.e. $\alpha_{pt} = 1$), which is also confirmed by practical measurements, is extremely advantageous for an efficient buyer's service processing as the probability grows with increasing α_{wt} (i.e. a decreasing level of burstiness). Furthermore, Fig. 10 points the obtained results when both gap processes have different mean values. Comparing the results drawn in Fig. 10 with the results depicted in Fig. 9 shows that in bursty situations, the mean values $m_1^{(wt)}$ and $m_1^{(pt)}$ have a minor influence on the buyer's service.

6 Conclusions

The theoretical findings of the present research are revealed by the establishment of the inter-relations between the queues, queue systems, queue models and service process. Also, the theoretical finding is the relationships between the terms 'customer', 'client', 'visitor', 'lost customer' and 'buyer'.

A proper service process can be seen as a key driver when analysing customer satisfaction. In this work, the focus was put on the analysis of buyers' service at the cash register by taking the unity of the waiting in the queue to the cash register and the buyer's payment processing into account. By applying the concept of generating functions to bursty business processes, a new model based on gap processes was introduced. The proposed model allows an analytical description of the service process at the cash register.

The practical application of the model of the service process as the unity of the waiting in the queue to the cash register and the payment processing based on the gap processes allows concluding that a given level of buyers' burstiness influences the buyers' service at the cash register. Further on, buyers' burstiness is a limiting parameter when analysing buyers' service. The model was confirmed by the data from two grocery shops in Lithuania.

Future work will focus on taking different means of payment (e.g. cash or credit card payment) as well as different number of available cash registers in parallel into account. Furthermore, the model of the service process considered as the unity of the waiting in the queue to the cash register and the buyer's payment processing at the cash register will be empirically validated.

References

1. C. Kiataramkul, K. Neamprem, Simulation of queueing system for vommercial bank in university: Case study of Bangkok Bank at King Mongkut's University of Technology North Bangkok, in *2019 Research, Invention, and Innovation Congress (RI2C)*, (2019), pp. 1–4. https://doi.org/10.1109/RI2C48728.2019.8999935
2. M.L.F. Cheong, Y.Q. Chia, Simulation model to evaluate effectiveness of queue management tool in supermarket retail chain, in *2019 IEEE International Conference on Industrial Engineering and Engineering Management (IEEM)*, (2019), pp. 606–610. https://doi.org/10.1109/IEEM44572.2019.8978794

3. A. Ahrens, A new digital channel model suitable for the simulation and evaluation of channel error effects, in *Colloquium on Speech Coding Algorithms for Radio Channels*, (London, 2000)
4. R. Adams, Active Queue Management: A Survey. IEEE Commun. Surveys Tutorials **15**(3), 1425–1476 (2013). https://doi.org/10.1109/SURV.2012.082212.00018
5. R. Hassin, M. Haviv, *To Queue or Not to Queue: Equilibrium Behavior in Queueing Systems* (Springer, Boston/Dordrecht/London, 2003)
6. L. Brink, R. Shorten, J.Y. Yu, Signaled queueing, in *Proceedings of the 2015 International Conference on Autonomous Agents and Multiagent Systems*, (AAMAS '15, International Foundation for Autonomous Agents and Multiagent Systems, Richland, 2015), pp. 811–819
7. D. Miceviciene, O. Purvinis, R. Glinskiene, A. Tautkus, Alternative solution for client service management. Appl. Res. Stud. Pract. **14**(1), 47–51 (2018)
8. R. Sharma, Mathematical analysis of queue with phase service: An overview. Adv. Oper. Res. 2014 (2014). https://doi.org/10.1155/2014/240926
9. S.C. Ferrari, R. Morabito, Application of queueing models with abandonment for Call Center congestion analysis. Gestao & Producao 27 (2020)
10. D. Hartleb, A. Ahrens, O. Purvinis, J. Zăšcerinska, Analysis of Free Time Intervals between Buyers at Cash Register using Generating Functions, in *International Conference on Pervasive and Embedded Computing and Communication Systems (PECCS). Online Streaming*, (2020)
11. Y. Tang, P. Guo, Y. Wang, Equilibrium queueing strategies of two types of customers in a two-server queue. Oper. Res. Lett. **46**(1), 99–102 (2018). https://doi.org/10.1016/j.orl.2017.11.009. http://www.sciencedirect.com/science/article/pii/S0167637717303243
12. D. Hartleb, A. Ahrens, O. Purvinis, J. Zăšcerinska, D. Micevǐciene, Internal and external factor analysis in bottleneck detection in shop sales: The case of grocery shops in Lithuania, in *International Conference on Pervasive and Embedded Computing and Communication Systems (PECCS). Online Streaming*, (2020)
13. C. Tonui Benard, C. Langat Reuben, J. Gichengo, On Markovian queuing models. Int. J. Sci. Res. (IJSR) **3**(11), 93–96 (2019)
14. G. Choudhury, M. Deka, A single server queueing system with two phases of service subject to server breakdown and Bernoulli vacation. Appl. Math. Model. **36**(12), 6050–6060 (2012). https://doi.org/10.1016/j.apm.2012.01.047. http://www.sciencedirect.com/science/article/pii/S0307904X12000625
15. H. Wilhelm, *Calculation of Error Structures in Binary Channels with Memory* (Books on Demand, Norderstedt, 2018)
16. J. Zhang, F. Shen, Y. Waguespack, Incorporating generating functions to com-putational science education, in *International Conference on Computational Science and Computational Intelligence (CSCI)*, (Las Vegas, 2016), pp. 315–320
17. S. Li, Generating function approach for discrete queueing analysis with decomposable arrival and service Markov Chains, in *IEEE International Conference on Computer Communications (INFOCOM)*, (Florence, 1992), pp. 2168–2177
18. H. Wilhelm, *Datenübertragung (in German)* (Milit¨arverlag, Berlin, 1976)
19. K.I. Goh, A.L. Barab'asi, Burstiness and memory in complex systems. Exploring Front. Phys. (EPL) **81**(4), 48002 (2008). https://doi.org/10.1209/0295-5075/81/48002

Adaptive, Automatic and Non-invasive Cultural Heritage Preventive Conservation Framework Based on Visual Information Crowdsourcing

Miguel Antonio Barbero-Álvarez, Juan Antonio Rodrigo, and José Manuel Menéndez

Abstract Cultural heritage pieces are prone to degradation inducted by ambient conditions. When happening, it is necessary to launch a maintenance process, which may be costly in terms of human means and/or money. In this paper we present the outline of a mean of practicing preventive conservation by monitoring the possible degradation by taking crowdsourced visual cues. We developed a special non-invasive display consisting in a colour calibration chart and indicating metal tags whose superficial colour change will act as cues of degrading ambiances. By crowdsourcing cell phone pictures of the display taken by third parties that go visit the heritage pieces, the conservation state of the piece can be monitored along time so damage can be minimized. In order to obtain valid crowdsourced information, we developed a self-designed colour calibration scheme, specifically designed for this work's context. We compare our calibration scheme with commercial solutions and state the necessity of developing specific solutions for projects where the requirements make the implementation of general means difficult. Our approach returns valid homogeneous colour information for the cultural heritage monitoring and surpasses other state-of-the-art works.

Keywords Cultural heritage conservation · Crowdsourcing · Colour calibration · Image processing

1 Introduction

Recently, people and government bodies have become aware of the importance tied to cultural heritage, and therefore a growing tendency concerning its preservation and care has appeared. On a technological and an engineering level, this has meant

M. A. Barbero-Álvarez (✉) · J. A. Rodrigo · J. M. Menéndez
GATV-SSR, Universidad Politécnica de Madrid, Madrid, Spain
e-mail: mba@gatv.ssr.upm.es

C.-C. Wang, A. Nallanathan (eds.), *Proceedings of the 5th International Conference on Signal Processing and Information Communications*, Signals and Communication Technology, https://doi.org/10.1007/978-3-031-13181-3_9

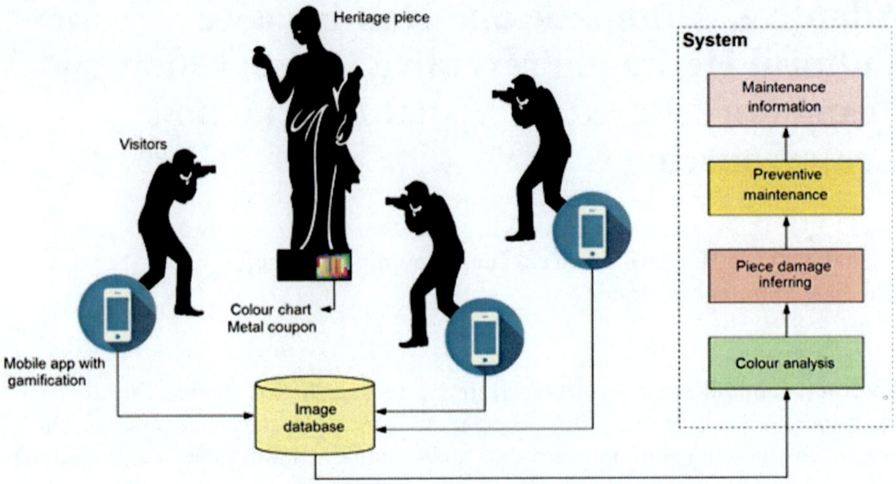

Fig. 1 Scheme of the proposed system

an appearance of a distinct field concerned in covering many aspects related to conservation employing digital devices [1].

Out of these, the preventive conservation is an aspect that seeks detecting possible damage or degradation in cultural heritage objects in time, in order to avoid future destruction or costly mending processes. Among the solutions proposed in this aspect, there exists a branch consisting in placing sensors to evaluate the physical conservation conditions of the pieces along time, and the information gathered from them treated accordingly to distinguish the exact moment where conservation needs to be launched [2].

With this paper we present a novel framework destined to enable preventive conservation of cultural heritage pieces exposed in museums by crowdsourcing visual information (Fig. 1). Visitors to the exhibitions are encouraged to take pictures with their cell phones of a special display placed next to the pieces and upload them via a specifically designed mobile phone app. The display contains a colour calibration charts and a set of material tags made of copper, silver and PH paper. These will change their surface colour depending on the ambient conditions [3], such acidity or humidity, which is perilous to the exposed pieces.

Therefore, pictures depicting these tags act as the temporal cues for surveilling the state of the exposed pieces. Museum visitors are encouraged to take photos using gamification techniques, offering prizes and discounts to encourage participation. In order to overcome the source camera disparity, given that no camera perceives all colours exactly the same way as another, we designed an adaptive, quick and effective colour calibration system that homogenises the crowdsourced sRGB information for its adequate process regardless of the origin [4, 5].

In the following sections, we describe the colour chart and the calibration proposals to be used in this system in detail and prove their effectiveness and results, surpassing related state-of-the-art works and commercial solutions. Our system, requiring only a small physical display to work, also has the advantage of reducing the presence of external monitoring elements that may invade the exhibition of the heritage pieces, given that both the information sources and processing are remote. It also avoids the need of installing dedicated sensors human surveillance efforts, reducing money costs in this direction too. In addition to this, the app employed to save the pictures also instructs third parties on how to correctly take images of the display, so an adequate distance, angle and focus that permit relevant information extraction can be assured. This is something an external surveillance monitoring by security cameras would not achieve, since their scope is wider and far away from ay exposed details.

Considering that commercial solutions for calibration purposes require human interaction and our system is meant to be both remote and automatic, these have been adapted by us to match the requirements of the system, and so the comparison with our proposals is made on the same ground.

2 Related Work

Crowdsourcing is a powerful manner to gather many kinds of data for scientific purposes [6], also involved in cultural heritage preservation [7], with multiple approaches and successful results [2, 8].

One of the ways of crowdsourcing data is having a constant feed of third-party sources, such as cell phones. This has the advantage of not needing to deploy a specific sourcing device, which is convenient for cultural heritage monitoring [2, 9].

Colour calibration addresses the problem on how different kinds of illuminants affect how colours are perceived, especially in the colour charts used for the process [10, 11].

Professional and commercially available proposals on calibration solutions may include both charts and mathematical basis [12–14] or an UI that requires manual operation [15].

There exist other engineering works that use the mathematical basis of publicly available calibration schemes, such as colour calibration matrix (CCM) [16], and modify it so it gets adapted to the working conditions. Some of them involve linear algebra approaches [17], least squares [18] or deep learning [19].

Other works design and develop their own colour charts matching the desired contexts [20].

As it can be seen, there exist solutions for the individual factors described, but there is no proposal that deals with all of them at the same time without being generalist or requiring human interaction. Our proposal, which will be described in the following sections, deals with the problem from scratch whilst designed for the specific purposes sought and in a totally automatic and adaptive way.

3 Materials and Methods

The crucial part of the monitoring system is the colour calibration, which needs to return reliable information for a good operation of the whole. In order to perform it, an adequate colour reference is needed to be represented in the pictures to be gathered. Furthermore, the mathematical basis of the calibration must be solid and needs to be properly designed, so the adaptive nature of the system for every picture can be achieved.

3.1 Colour Chart Design

We designed a colour chart that implies mathematical regularity. Given that the sRGB colour space takes form of a Cartesian three-axis space, each dimension is divided in regular segments, so each combination of them forms the 'anchor colours', or the particular colorimetric values to be displayed in the charts as patches. The amount of colours is decided upon the number of patches that can be deployed and also based on an acceptable physical size of the chart in order to avoid invasiveness towards the piece to be surveilled, which is one of the requirements of this project.

In Eq. (1) *n_bits* is the number of quantification bits by colour.

Note that the root term should return an integer, so the division of the axes is not fractional.

We adjusted the amount of patches to 64. Then, the physical chart is big enough to display 8-bit colour anchors of a decent size to extract their digital values from the pictures without problem but small enough to not mean an excessive invasiveness that may hinder the exhibition of the piece.

The featured selection includes the limits of the sRGB space, primaries and black and white (Fig. 2, left).

However, mathematically ideal colours as the sRGB and CYM primaries, or pure black or pure white, are not likely to appear in digitally acquired images of real scenes. Normally, they feature a limited gamut, spanning the interior of the total range of perceivable colours in sRGB. Therefore, there is a need to depict 'real' values out of the preconceived division of the space as the wanted colour anchors, so a reasonable and accurate calibration process is performed.

In order to reach this reality, a preliminary chart depicting the ideal sRGB division as the colour patches has been printed, and then, the spectral values measured with a Konica Minolta CM-700-d spectrophotometer. These are then transformed to sRGB afterwards (Fig. 2, right, and Fig. 3), showing the real counterpart of the ideally conceived anchors.

Thus, a sensible and realistic gamut for the calibration chart is achieved.

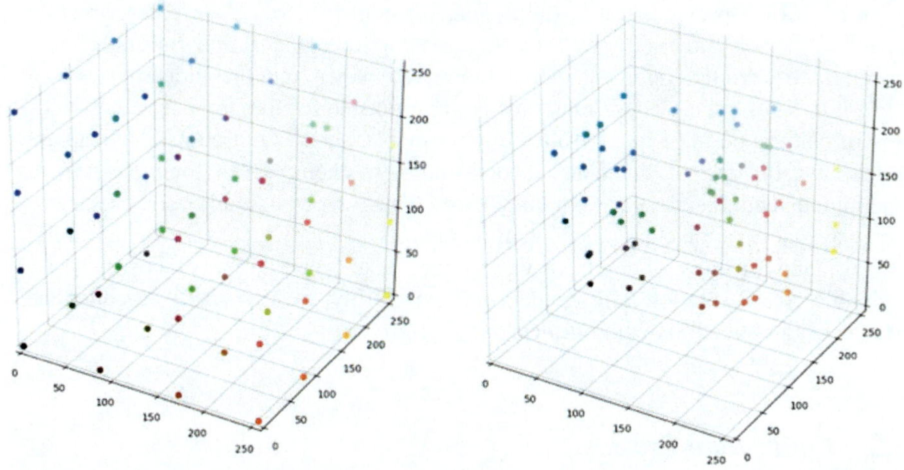

Fig. 2 Ideal partition of the sRGB space (left); the same partition with the real counterparts (right)

Fig. 3 Colour chart employing the original division of the sRGB space (left); the same colour chart after calculating the real-life counterpart values (right)

The printing process by its own means an inevitable reduction of the span of the resulting gamut represented by the anchor colours. But, if we consider the full sRGB range in the preliminary division, the post-printing reduction will be less affected. If we, on the other hand, decide that only an arbitrary interior of the sRGB space were to be considered for building the anchor colours from at the preliminary division, its span reduction in space, when printed, would mean that an even smaller region would be resulting afterwards. So, an arbitrarily considerable region of space near the limits of the sRGB cube would be left outside of the calibration process, thus losing much more colour information to be calibrated.

In the end, the 64 patches have been physically arranged in 11 columns and 9 rows. A space of seven columns and five rows is left free in the middle. It is reserved for placing tags made of metal or paper when deploying the system. These will change their surface colour accordingly to the possible degradation, which will be the informational cues for the maintenance system (Fig. 3).

In one of the previous publications related to our project [21], we described the selection process and criteria we took for choosing the physical materials the chart is meant to be printed upon: acid-free laminated paper and the industrial ink set LED-UV Xtreme Pro. These materials are appropriate for the preventive conservation of cultural heritage in controlled environments, such as museums. A cheap and quick reproduction of the chart is possible with them, given their easiness to manipulate and their reduced money cost. This favours simultaneous usage of many instances of the chart placed in diverse spots, so many pieces can be monitored at once and pictures of them crowdsourced. The bigger prize of commercially available charts and solutions hinders this possibility, making them less adequate to cover big collections with our designed system.

3.2 Calibration System

Analogously to our decision on designing the chart, we developed some self-designed calibration approaches that are suited for the goals followed with our project. When considering that the source of each crowdsourced picture can be seen as unique, it is mandatory that the calibration technique should be adaptive while suited for the digital domain. Also, the possibility of obtaining a big quantity of crowdsourced data in a short span of time forces the process to be quick and efficient, so, at least, all data gathered in a day can be processed within the same period.

Therefore, two mathematically generalist calibration techniques have been developed that match the aforementioned requirements. Based on value interpolation between given anchors, the first developed approach treats every sRGB channel separately, and the second one takes information from all dimensions to calculate each single resulting coordinate. Thus, the interpolated curves act as transfer functions that translate the original crowdsourced sRGB values to the calibrated space we need to obtain for each cue.

In addition, we have compared the results obtained by both techniques with a commercial calibration solution, such as the colour calibration matrix (CCM). Nevertheless, since most of the times this solution is implemented in external software that requires human interoperability, we programmed it by ourselves so it becomes automatic and matches the project requirements of automatic processing. We followed the theory outlined in [17].

The starting point for all three compared calibration techniques consists in extracting the sRGB values of the colour patches displayed in the gathered pictures and associate them to the ideal colour anchors, so the 'skeleton' of the transfer functions can be built. This is achieved with Reinhard's colour transfer formula [22], so the colour distributions of each patch in the pictures are spatially translated in each dimension.

From this point on, each calibration approach differs. Firstly, for both the one-dimensional and multidimensional calibration, the connection between each point in the 'skeleton' is calculated via linear interpolation. This way, three multi-sloped curves (one for each colour channel) and three four-dimensional hyper-surfaces are generated for the former and the latter approach, respectively.

The expression in Eq. 1 states how to obtain the Y coordinate for a given point y_p when knowing the interpolated points (x_1, y_1) and (x_2, y_2) and its X coordinate x_p. This approach is followed to build the three transfer functions in the one-dimensional calibration process, one for the R channel and the others for G and B, using the single coordinates of the colour patches. After having built them, every pixel of the image is treated by processing each of its coordinates with the corresponding function.

$$y_p = y_1 + (x_p - x_1)\frac{y_2 - y_1}{x_2 - x_1} \tag{1}$$

$$f(x_1, x_2, x_3) = \sum\nolimits_{k=1,2} P(r_k, g_k, b_k) \prod\nolimits_{i=1}^{3} \begin{cases} x_1 & k = 2 \\ (1 - x_1) & k = 1 \end{cases} \tag{2}$$

Therefore, the R, G and B channels are separately treated in each picture and then joined back, making the calibrated result.

The multidimensional approach bases itself on the former but expands it to three dimensions, calculating each of the calibrated R, G and B layers using information from all three channels at the same time in every picture. The calculated hyper-surfaces $f(x_1, x_2, x_3)$ are thus based upon a trilinear interpolation (Eq. 2). For each point in space to be calculated, there is a surrounding 'cube' $P(r_k, g_k, b_k)$, $k = \{1,2\}$ made of the immediately next sRGB anchor points that serve as the ground for the interpolation in the vicinity.

After building the hyper-surfaces for the image to be treated in this way, and similarly to the former, every pixel is projected into the calibrated space.

Both techniques perform automatically and adjust their calculations adaptively based on the data every single gathered image displays, compared with the established anchor colours. Our adaption of CCM here matches the same requirements, so no human worker is needed to perform the maintenance, which would not be possible if purchasing the commercially available version.

In order to deduce the CCM, representative values of each patch (such as the average) are compared with the anchor colours, and a 3×3 matrix is deduced (Eq. 4).

$$R = \text{CCM} \times O_r \tag{3}$$

$$\text{CCM} = \left(R \times O_r^T\right) \times \left(O_r \times O_r^T\right)^{-1} \tag{4}$$

The set of anchor colours R should be equivalent to the calibration matrix multiplied by the patch representative values of each image Or (Equation3). Then, the CCM can be deduced with Eq. 4.

The calibrated picture is obtained by multiplying the *R*, *G* and *B* coordinates of each pixel with the obtained matrix.

Calibrating following this approach saves colour distribution information within the picture, given that it assigns a different target value to every original value.

Nevertheless, care needs to be taken, so the gamma non-linearity of the sRGB space does not affect the results of a linear operation such as the matrix multiplication. Therefore, this CCM approach requires a gamma process before and after the multiplication, so the sRGB pixel information remains coherent.

4 Experimental

The proposed designs have been tested both in lab and in field. Firstly, the lab tests have been performed over charts with 'dummy' tags printed on the free space in the middle. These two sets of 'dummy' tags depict real textures taken by photographs of silver, copper and PH paper and uniform colours of hues found within them. Then, the printed charts have been photographed with commercial brands of cell phones under different lighting temperatures and intensities. It is natural to assume that, when exposing different instances of the same object in different locations, the lighting may be different for each one. So the lab tests take the crowdsourced camera uncertainty and the illumination disparity into account to test the reliability of the calibration techniques.

Overall, five sets of four to six pictures have been taken in the lab, under low, medium and high colour temperatures and low and high illumination intensities (Fig. 4). Each one of these has been calibrated with the three outlined techniques and the results saved for their comparison.

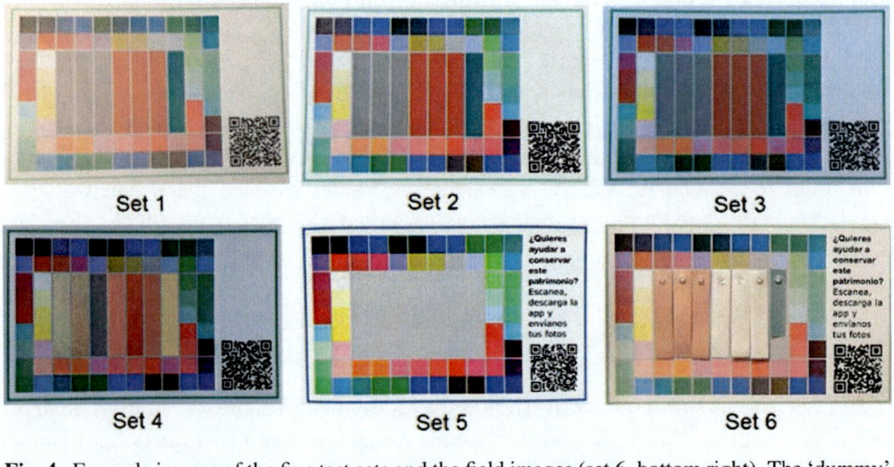

Fig. 4 Example images of the five test sets and the field images (set 6, bottom right). The 'dummy' dosimeters are to be noted in sets 1–4 and their real-life counterparts in set 6

The test results have been evaluated transforming the data to the CIELAB colour space, which is specifically designed for measurements and quality works. For each calibrated picture, the CIELAB data from the colour patches have been extracted and compared with the original anchor colours. With this, the accuracy of the built 'skeleton' of the transfer functions can be evaluated.

Additionally, the same has been performed with the 'dummy' tags, so the ability of the calibration to accurately correct the most sensitive regions of the space, where the indicating tags reside.

For this, the CIEDE2000 colour difference has been employed that acts as a reliable distance in CIELAB space, widely used in the evaluation field.

After having assessed the results of the lab tests, a preliminary version of the automatic preventive conservation system has been deployed in some museums (Museo Nacional de Ciencia y Tecnología – MUNCYT) in Madrid and A Coruña, Spain. For a 1-year trial period, real crowdsourced pictures will be gathered and employed as field cases to improve the system. Therefore, the performance of the calibrations is also assessed with these (Fig. 4, bottom right).

Nevertheless, given that the real crowdsourced pictures will depict the real indicative tags, there exists no 'reference' to be compared with. Consequently, only the patches have been evaluated in the aforementioned manner. The results of six of the field images are explained in the following section.

5 Results and Discussion

The following Tables 1 and 2 outline the results explained in the previous section.

Inspecting the results coming from comparing the three calibration procedures when building the 'skeleton' of the calibrated space, it is noteworthy to remark that the multidimensional calibration offers the best results. The CIEDE2000 difference between the calibrated patches and the original references stays between 0 and 1, as seen in Table 1, which implies an extremely accurate value, also revealed by the low standard deviation values. The linear calibration offers some variability in their values but mostly stays within the 0–10 difference range, with some infrequent outliers of a bigger magnitude. Nevertheless, even with this kind of results, the linear calibration still surpasses other state of art works of similar focus that employ commercial calibration solutions.

The CCM calibration is seen to not be suited for this project. The extreme variability and the bigger magnitude of the CIEDE2000 differences for each patch are vastly surpassed by the other two.

When visually inspecting the resulting pictures, it can be confirmed that the multidimensional approach surpasses the effects of the variable illumination conditions and camera characteristics better than the one-dimensional one. The CCM provokes an exaggerated overall colour shifting depending on the illumination temperature of the original picture that even varies from different images in the same context. Therefore the usage of CCM is discarded from this project. The same conclusions are reached both in the test and the field images.

Table 1 CIEDE2000 means and standard deviations in the 64 patches for the experiments employing all three calibration approaches

	Multidimensional		Linear		CCM			Multidimensional		Linear		CCM	
Patch	Mean	Std.	Mean	Std.	Mean	Std.	Patch	Mean	Std.	Mean	Std.	Mean	Std.
0	0.41	0.15	14.08	3.05	16.76	6.35	32	0.19	0.06	5.09	2.02	14.38	7.39
1	0.26	0.09	9.21	3.59	18.69	3.47	33	0.19	0.08	3.44	1.65	14.63	4.03
2	0.22	0.07	8.10	4.50	15.61	4.07	34	0.23	0.05	7.05	2.87	16.77	5.49
3	0.30	0.12	9.91	4.52	27.20	5.29	35	0.20	0.16	9.47	5.17	10.28	3.36
4	0.17	0.07	5.02	3.10	13.49	6.64	36	0.21	0.09	6.64	3.23	16.39	6.62
5	0.19	0.08	8.93	3.66	27.23	5.06	37	0.19	0.07	6.56	2.37	15.27	5.78
6	0.20	0.03	4.93	1.74	15.04	5.90	38	0.21	0.06	4.93	4.71	11.77	4.03
7	0.17	0.06	7.40	2.36	17.79	4.58	39	0.37	0.42	9.04	5.17	9.30	3.61
8	0.30	0.05	6.85	5.47	13.77	2.22	40	0.25	0.08	4.93	1.38	16.53	4.73
9	0.29	0.06	10.37	3.53	13.92	2.23	41	0.23	0.10	5.06	2.07	14.72	4.99
10	0.26	0.09	7.05	2.22	12.90	3.70	42	0.30	0.11	4.77	2.44	11.95	4.26
11	0.58	0.82	5.53	1.41	20.19	3.55	43	0.25	0.11	5.55	3.16	9.75	3.87
12	0.18	0.05	3.70	1.97	14.57	7.29	44	0.20	0.06	4.41	2.14	13.11	6.06
13	0.21	0.08	7.57	1.85	25.05	5.27	45	0.20	0.05	4.34	2.13	12.46	6.74
14	0.18	0.06	4.05	1.58	14.01	7.58	46	0.14	0.06	2.69	1.71	13.40	5.55
15	0.15	0.07	4.24	2.75	13.61	6.19	47	0.22	0.09	3.63	2.74	13.26	3.65
16	0.18	0.09	5.40	2.64	15.45	6.68	48	0.17	0.06	4.19	1.13	15.83	2.25
17	0.18	0.06	4.66	3.15	12.19	4.49	49	0.22	0.08	3.99	0.74	15.68	2.37
18	0.37	0.17	13.23	4.99	14.37	5.86	50	0.19	0.08	3.45	2.00	13.22	2.68
19	1.54	1.32	11.05	4.73	23.15	3.77	51	0.22	0.07	5.65	2.80	12.16	4.45
20	0.19	0.11	8.82	3.72	20.47	5.06	52	0.21	0.07	2.97	1.49	14.46	2.83
21	0.23	0.09	9.85	3.41	23.65	5.36	53	0.20	0.07	3.09	1.27	13.34	3.47
22	0.16	0.05	4.91	2.53	15.61	6.22	54	0.20	0.05	5.36	1.79	13.96	2.95
23	0.17	0.07	6.42	3.86	22.85	5.91	55	0.19	0.05	7.59	3.54	12.84	4.79
24	0.18	0.04	4.99	1.30	15.82	5.42	56	0.13	0.04	4.39	2.11	17.28	2.06

25	0.22	0.07	3.54	1.68	16.79	3.75	57	0.18	0.06	3.88	2.10	15.15	2.94
26	0.13	0.08	8.33	3.08	11.21	3.62	58	0.16	0.04	5.10	2.09	17.61	5.15
27	0.48	0.69	4.32	1.06	21.73	3.79	59	0.19	0.04	6.18	1.56	17.16	5.63
28	0.50	0.16	10.71	1.97	12.76	3.60	60	0.13	0.06	2.10	0.73	18.94	5.22
29	1.66	1.72	6.81	2.90	22.49	3.58	61	0.13	0.07	2.82	1.58	18.16	5.51
30	0.23	0.03	4.43	2.37	15.98	7.17	62	0.20	0.10	3.93	1.19	16.07	6.53
31	0.19	0.07	5.60	2.24	18.27	3.82	63	0.19	0.18	5.10	1.98	15.25	5.32

Table 2 CIEDE2000 means for the sets of 'dummy' tags for all experiments in the three calibration schemes

Uniform tag	Mean CCM	Mean multidimensional	Mean one-dimensional
0	14.08	6.79	11.87
1	13.87	6.62	10.15
2	13.45	6.38	9.73
3	16.29	9.71	9.85
4	15.83	9.49	9.88
5	15.43	9.45	11.98
6	32.55	5.32	17.57
Textured tag	**Mean CCM**	**Mean multidimensional**	**Mean one-dimensional**
0	13.31	4.51	3.92
1	14.03	8.81	6.51
2	13.03	7.71	10.37
3	17.92	7.77	5.72
4	15.35	8.87	10.33
5	16.02	9.83	6.25
6	15.76	9.48	7.81

When inspecting the 'dummy' tags in the test images, similar conclusions regarding the calibration approaches can be taken. Table 2 reveals the lower difference and more stable value range of the multidimensional calibration. Nevertheless, considering some of them feature textures with bigger value variability than the uniform colour patches, especially in the one-dimensional proposal, it is also interesting to confirm how the process may exaggerate it or not. Thus, standard deviations of the dosimeters after the calibrations have been taken.

The multidimensional process overcomes the variability, offering a smaller standard deviation. The linear approach produces some pixelated effects that assign false values due to the independent processing of each one of the three coordinates, so errors here are more prone to happen. CCM, on the other hand, respects the original pixel distribution given that the matrix multiplication is a linear process, but the other aforementioned defects make it non-viable.

6 Conclusions

In this paper we outlined the importance of the colour chart design and mathematical basis of colour calibration for a cultural heritage preventive maintenance system based on crowdsourced pictures. We explained the reasons why commercial solutions are not suited for this working context and the necessity of developing self-designed resources, namely, related to cost and difficulty to deploy in a massive scale.

The results we present show with objective measures the excellent quality of our approaches. We demonstrated that both calibration techniques we propose surpass the results of other state-of-the-art works of similar context such as [9], and adapting commercial solutions trying to overcome the aforementioned obstacles does not necessarily imply it will adequately work. The good functioning of our proposal has been demonstrated in both simulated and real cases that feature the adverse conditions expected to be found in a picture crowdsourcing setting, as illumination and acquisition disparities.

Our solution is adaptive from scratch and works in an all-digital environment, thought for sRGB and 8-bit information, as most commercial phone cameras work. Aside, the required gear (a relatively small calibration chart to be placed near the pieces under surveillance), along with the absence of permanent physical acquisition devices, minimizes the invasiveness of the system, which is a mandatory requirement to fulfil.

To conclude, we state that developing a self-designed framework and assets to only fulfil specific requirements and goal is a feasible option in an engineering project that can offer decent results and leaves place for further flexibility, improvements and changes when the mandatory conditions hinder the usage of commercial solutions.

In addition, the involvement of visitors in the preventive conservation of the pieces, through gamification techniques, can be a very important element in making visitors aware of the need to preserve our heritage.

Acknowledgement We want to thank Emilio Cano and Blanca Ramírez-Barat from CENIM-CSIC for their material contributions and supervision of the project.

This work was supported in part by a grant from the project MIPAC-CM (Monitorización por procesado de imagen y ciencia ciudadana para la conservación de materiales del patrimonio cultural – Monitoring by image processing and citizen science for conservation of cultural heritage materials), project code Y2018/NMT-4913.

Author Contributions Miguel Antonio Barbero-Álvarez designed the calibration schemes, performed experiments, curated data, corrected errors and drafted the paper.

Juan Antonio Rodrigo designed the chart, curated data and supervised the experiments and drafting.

José Manuel Menéndez conducted the research, curated data, supervised the designs and experiments and corrected drafting.

References

1. S. Battiato, G.M. Farinella, F. Milotta, A. Ortis, L. Addesso, A. Casella, V. D'Amico, G. Torrisi, The social picture, in *Proc. ACM Int. Conf. Multimedia Retr*, (2016), pp. 397–400
2. E. Angelini, F. Civita, S. Corbellini, D. Fulginiti, A. Giovagnoli, S. Grassini, M. Parvis, Innovative monitoring campaign of the environmental conditions of the stibbert museum in florence. Appl. Phys. A **122**, 02 (2016)

3. P. Han, D. Dong, X. Zhao, L. Jiao, Y. Lang, A smartphone-based soil color sensor: For soil type classification. Comput. Electron. Agricult. **123**, 232–241 (2016)
4. M.A. Barbero-Álvarez, J.M. Menéndez, J.A. Rodrigo, An adaptive colour calibration for crowdsourced images in heritage preservation science. IEEE Access **8**, 185 093–185 111 (2020)
5. M.A. Barbero-Álvarez, J.A. Rodrigo, J.M. Menéndez, Self-designed colour chart and a multi-dimensional calibration approach for cultural heritage preventive preservation. IEEE Access **9**, 138371–138384 (2021). https://doi.org/10.1109/ACCESS.2021.3119223
6. M. Haklay, *Citizen Science and Volunteered Geographic Information: Overview and Typology of Participation* (Springer, Dordrecht, 2013), pp. 105–122. https://doi.org/10.1007/978-94-007-4587-2_7
7. K. Makantasis, A. Doulamis, N. Doulamis, M. Ioannides, In the wild image retrieval and clustering for 3D cultural heritage landmarks reconstruction. Multimedia Tools Appl. **75**(7), 3593–3629 (Aug. 2014)
8. C. Daffara, G. Marchioro, D. Ambrosini, Smartphone diagnostics for cultural heritage. Proc. SPIE **11058**, 110581K (2019)
9. R. Brigham, J. Grau-Bové, A. Rudnicka, M. Cassar, M. Strlic, Crowdsourcing as an analytical method: Metrology of smartphone measurements in heritage science. Angew. Chem. **130**(25), 7545–7549 (Jun. 2018)
10. P. Marrero Fernandez, F.A. Guerrero Peña, T. Ing Ren, J. Leandro, Fast and robust multiple colorchecker detection using deep convolutional neural networks. Image Vis. Comput. **81**, 11 (2018)
11. C.-L. Chen, S.-H. Lin, Intelligent color temperature estimation using fuzzy neural network with application to automatic white balance, in *2010 IEEE International Conference on Systems, Man and Cybernetics*, (2010), pp. 796–803
12. XRite, i1publish. Last access: April 2021, https://www.xrite.com/categories/calibration-profiling/i1publish
13. XRite, Colorimetric values for ColorChecker Family of Targets. Last access: April 2021, https://xritephoto.com/colorcheckerkits
14. Imatest, Imatest products. Last access: April 2021, https://www.imatest.com/products/
15. ImageJ, Fiji. Last access: April 2021, https://imagej.net/Fiji
16. Imatest, Color Correction Matrix (CCM). Last access: April 2021, https://www.imatest.com/docs/colormatrix/
17. S. Sunoj, C. Igathinathane, N. Saliendra, J. Hendrickson, D. Archer, Color calibration of digital images for agriculture and other applications. ISPRS J. Photogr. Remote Sensing **146**, 221–234 (2018) [Online]. Available: https://www.sciencedirect.com/science/article/pii/S0924271618302600
18. J. Vaillant, A. Clouet, D. Alleysson, Color correction matrix for sparse RGB-W image sensor without IR cutoff filter, in *Unconventional Optical Imaging*, (SPIE, Strasbourg, 2018), p. 3. [Online]. Available: https://hal-cea.archives-ouvertes.fr/cea-02390508
19. A. Abdalla, H. Cen, E. Abdel-Rahman, L. Wan, L. He, Color calibration of proximal sensing rgb images of oilseed rape canopy via deep learning combined with k-means algorithm. Remote Sensing **11**, 3001 (2019)
20. Y. Motonaga, T. Matsumoto, N. Motonaga, Color chart of european pear 'le lectier' based on the color image analysis, in *Proceedings of SICE Annual Conference 2010*, (2010), pp. 2455–2461
21. B. Ramírez Barat, E. Cano, M. Molina, M.A. Barbero-Álvarez, J.M. Menéndez, J.A. Rodrigo, Design and validation of tailored colour reference charts for monitoring cultural heritage degradation. Heritage Sci. **9**(41) (2021)
22. E. Reinhard, M. Adhikhmin, B. Gooch, P. Shirley, Color transfer between images. IEEE Comput. Graph. Appl. **21**(5), 34–41 (2001)

An Inverted-ITL Algorithm for Mining Partial Periodic-Frequent Patterns

Ye-In Chang, Xin-Long Chen, and Sheng-Hsin Chiang

Abstract In this paper, we propose an *Inverted-ITL* algorithm for mining partial periodic-frequent patterns. Although the *GPF-growth* algorithm has been proposed to achieve the same goal, it needs to construct the prefix tree which needs too much storage space and processing time. Moreover, the *GPF-growth* algorithm will scan the database twice and sort each transactional list. To avoid those disadvantages, in this paper, we propose the *inverted-ITL* algorithm. Our algorithm is more efficient than the *GPF-growth* algorithm. We will use one data structure, the *ITL-tree*, to store the items which appear in the database. Therefore, our algorithm can need shorter processing time than the *GPF-growth* algorithm.

Keywords Frequent patterns · Itemsets mining · Periodic patterns · Periodic-frequent patterns · Transactional database

1 Introduction

In recent years, the research of data mining [1, 2] has become more popular. There are many techniques for mining interesting patterns in the database, like frequent pattern mining [3–5], frequent weighted pattern mining [6], frequent closed pattern mining [7], maximal frequent pattern mining [8], and periodic-frequent pattern mining [9–13]. Among the above techniques, current researches on periodic-frequent pattern mining have focused on discovering full periodic-frequent patterns in the database. However, partial periodic-frequent patterns are more common in the real world, i.e., finding frequent patterns which frequently occur but do not successively occur one after one in the database. The reason for this phenomenon is the imperfect nature of the real-world.

Y.-I. Chang (✉) · X.-L. Chen · S.-H. Chiang
Department of Computer Science and Engineering, National Sun Yat-Sen University, Kaohsiung, Taiwan
e-mail: changyi@mail.cse.nsysu.edu.tw

C.-C. Wang, A. Nallanathan (eds.), *Proceedings of the 5th International Conference on Signal Processing and Information Communications*, Signals and Communication Technology, https://doi.org/10.1007/978-3-031-13181-3_10

The periodic-frequent pattern mining is an extension of the frequent pattern mining. In the frequent pattern mining, the count of the support is the factor which we only care about. However, in the periodic-frequent pattern mining, we do not just care about the frequency of each item but also need to make sure that the frequent pattern occurs periodically in the transactional database [9]. According to the property of the transactional database, the database consists of two fields: (1) timestamp and (2) item. The timestamp denotes the current time, when a transaction list is generated. For the topic of mining periodic-frequent patterns, the item will not appear repeatedly in a transaction list. Moreover, we do not have to care about the number of the items in a transaction list. That is, a transaction list {*a*, *a*, *b*, *c*} will not appear. Moreover, {*a*, *b*, *c*} and {*b*, *c*, *a*} are the same. The timestamp is used to record the time, when a transaction list is generated and the transaction lists are sorted according to the ascending order of the timestamp. Therefore, for the entire database, there may exist time that no transaction list is generated. For instance, the list of transactions may contain [*T*1, *T*2, *T*3, *T*5, *T*6], where time *T*4 does not appear.

Take Table 1 as an example, which is denoted by transactional database TDB1. Assume that the user-specified minimum support is 2, and the user-specified maximum period is 3. At first, we scan all the transactions of transactional database TDB1 to count the total support and obtain the maximum period of the complete set of periods for each item. We can exploit these data items to discover which patterns are periodic-frequent patterns. Figure 1 shows the result after scanning database TDB1, which is denoted by database TDB2. In database TDB2, the symbols *ItemN*, *CountN*, and *MaxD* denote the item name, the count of the total support, and the maximum period of the complete set of periods for each item, respectively.

According to the data of transactional database TDB2, we can know the support and the maximum period of each item in the database. Take item *a* as an example. *CountN* is larger than or equal to 2, and *MaxD* is lower than or equal to 3. So item "a" is a periodic-frequent pattern. Let's focus on item "c," which *CountN* is not less than 2, but its *MaxD* is larger than 3, so item *c* is not a periodic-frequent pattern. Next, for item *i*, it is not a periodic-frequent pattern for the same reason as item *c*. Although its *CountN* is larger than 2, its *MaxD* is larger than 3. Then, items *j* and *g* are not periodic-frequent patterns, because they have the same reason as the items *c* and *i*. That is, the *MaxD* is larger than the threshold. Finally, item "h" also is not a periodic-frequent pattern. Because its *CountN* and *MaxD*, neither of them meet the threshold. Based on the definition of *anti-monotone*, pruning these items will not affect the result. Therefore, the result after the pruning step contains items *a*, *b*, *d*, *e*, and *f*.

Table 1 An example of the transactional database TDB1

ts	Items	ts	Items
1	*ab*	6	*degf*
2	*acdi*	7	*abi*
3	*cefij*	8	*cdej*
4	*abfgh*	9	*abef*
5	*bd*	10	*acgi*

ItemN	CountN	MaxD
√a	6	3
b	5	3
*c	4	5!
d	4	3
*i	4	4!
e	4	3
f	4	3
*j	2	5!
*g	3	4!
*h	1!!	6!

! : larger than 3

!! : lower than 2

* : not periodic-frequent pattern

Fig. 1 After scanning all transactions of database TDB1, called database TDB2 (the minimum support = 2, the maximum period = 3)

Next, we combine those items into the new candidate size 2 patterns. Then, after checking, the qualified patterns are *ab* and *ef*. In the same way, we combine the periodic-frequent size 2 patterns into the new candidate size 3 patterns to discover the periodic-frequent size 3 patterns. The similar process is repeated until no periodic-frequent patterns are generated.

Recently, Kiran et al. have proposed the *PFP-growth++* algorithm [10], a prefix tree-based algorithm. They use the prefix tree with the unique timestamp to mine the periodic-frequent patterns. However, they do not take the real-world situation into account. Because in their algorithm, as long as the pattern does not meet the threshold once, it will not be considered as the periodic-frequent pattern. Later, Kiran et al. proposed the *GPF-growth* algorithm [12], which is also a prefix tree-based algorithm. The *GPF-growth* algorithm [12] considers fault tolerance by adding parameter periodic ratio. It can effectively solve the problems encountered by the *PFP-growth++* algorithm [10].

The *GPF-growth* algorithm [12] can find more realistic patterns than the *PFP-growth++* algorithm [10]. The *GPF-growth* algorithm considers about the fault tolerance of each pattern. However, we think that the construction of the prefix tree of *GPF-growth* needs too much time and memory space because the *GPF-growth* algorithm will scan the database twice and reorder each transactional list. Moreover, it constructs multiple prefix trees. Therefore, in this paper, we propose the *Inverted-ITL* algorithm to reduce the processing time. In our algorithm, we just only scan database once and store information of each pattern in two data structures. Later, we use these structures to find the periodic-frequent patterns. Therefore, our algorithm can need less processing time than the *GPF-growth* algorithm. Note that we do not need too much time to sort each transactional list and construct multiple prefix trees. From our performance study, we show that the performance of our algorithm is more efficient than that of the *GPF-growth* algorithm.

2 Related Works

The Apriori algorithm [3] has a great start and contribution to the research of data mining with frequent patterns. However, the algorithm is hard to achieve good performance. Later, Han et al. propose the *FP-growth* algorithm [5] for mining frequent patterns which scans twice in the database, and the algorithm exploits a tree structure, called FP-tree with the count of each item in the FP-tree. Next, Deng et al. propose the *PrePost* algorithm [4] and the *dFIN* algorithm [4] for mining frequent patterns with the property of scanning the database twice. Moreover, the algorithm exploits a tree structure, called PPC-tree which is an extension of the FP-tree and a vertical data structure, called N-list for each item in the PPC-tree. There are many algorithms for data mining which are developed based on this structure, such as the *PFP-growth* algorithm [14], the *NAFCP* algorithm [7], the *INLA-MFP* algorithm [8], and the *NFWI* algorithm [6]. Although many algorithms have been proposed as mentioned above, none of them have considered the period. Kiran et al. have proposed the *PFP-growth++* algorithm [10] for mining periodic-frequent patterns, which considers both the frequency and the period for each item. Later, Kiran and Reddy use the simplified model [11] to find all frequent patterns which have exhibited complete cyclic repetitions in the database. Kiran et al. propose the *MCPF-model* [13] to discover periodic-frequent patterns involving both frequent and rare items effectively. Kiran et al. [9, 10] have discussed greedy search techniques to discover periodic-frequent patterns effectively. All of these researches have focused on finding full periodic-frequent patterns. Then, Kiran et al. propose the *GPF-growth* algorithm [12], which considers about the fault tolerance for each item. The whole algorithm is basically the same as the *PFP-growth++* algorithm. The difference is the way to decide whether a pattern is a periodic-frequent pattern, which will be determined according to the percentage of the interest period. The components that make up the *GPF-growth* algorithm are the GPF-list and the GPF-tree. The *GPF-growth* algorithm forms the GPF-list by scanning the database once. Then, the GPF-list is sorted in the descending order of support for each item. Finally, they recursively mine the GPF-tree to discover the complete set of partial periodic-frequent patterns. Therefore, the *GPF-growth* algorithm needs to scan the database twice, sort the items once and the database once, and build a tree structure to find a complete set of the partial periodic-frequent patterns.

3 The Inverted-ITL Algorithm

In this section, we will introduce our algorithm called the Inverted-ITL (Inverted-Item-Time-List) algorithm to mine the partial periodic-frequent patterns.

3.1 Data Structure

In this subsection, we will use an example database TDB3 to explain our algorithm. Note that there is no transaction with $ts = 5$. During the process of partial periodic-frequent pattern mining, we exploit the support, the interesting period, and the periodic-ratio as main factors to determine whether the pattern is the partial periodic-frequent pattern or not. Thus, we use the *ITL-tree* and the *ITL-list* to record the information for each item. Note that we only have to scan the database once to record the information and we do not need to do any sorting operation with the database. During the mining process, we will frequently use these data structures to find the partial periodic-frequent patterns. Table 2 shows an example database for timestamp *ts*, and Table 3 shows the variables used in our algorithm. Next, we will illustrate the two data structures which we will use them in our algorithm.

The *ITL-tree*, a tree structure, is used to store all the patterns which appear in the database. It is a prefix tree, where each node may become a partial periodic-frequent pattern, but the root is an empty pattern. All k-length patterns are stored at level k of the tree, where $k \geq 1$. For instance, in Fig. 2, square boxes represent noncandidate patterns, dotted circles represent the candidate patterns, and solid circles represent the partial periodic-frequent patterns. Moreover, in the *ITL-tree*, once the pattern is confirmed as a noncandidate pattern, its super-set pattern will not appear. In other words, no pattern will generate new patterns with noncandidate patterns. Finally, the patterns in the solid circle are the partial periodic-frequent patterns that we want to find.

Table 2 An example of the transactional database TDB3

ts	Items	ts	Items
1	*ac*	8	*cdfg*
2	*abg*	9	*ab*
3	*de*	10	*cdef*
4	*abcd*	11	*abcef*
6	*abcd*	12	*abcd*
7	*abe*	13	*cef*

Table 3 Variables

Variable	Definition
ItemN	The name of the pattern
CountN	The count of the pattern
RangeC	The number of the pattern which is not larger than MaxPer during the period
FirstT	The first timestamp of the pattern
LastT	The last timestamp of the pattern
BP	Representing the timestamp of the bit pattern
Mark	A Boolean flag to check whether the last period of the pattern is not larger than the MaxPer
MinSup	The minimum support threshold
MinPR	The minimum periodic-ratio threshold
MaxPer	The maximum period threshold

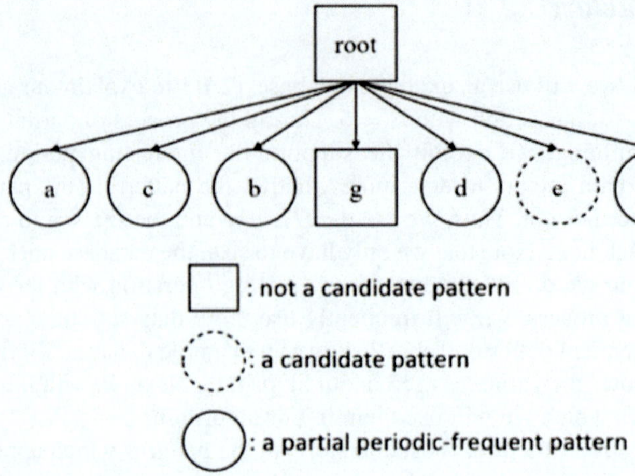

Fig. 2 The simple diagram of the *ITL-tree*

Fig. 3 The *ITL-list* and the *BP* table after scanning the second transaction (*abg*): (**a**) the *ITL-list* and (**b**) the *BP* table

ItemN	*a*	*c*	*b*	*g*
CountN	2	1	1	1
RangeC	2	1	1	1
FirstT	1	1	2	2
LastT	2	1	2	2

(a)

ItemN	T1	T2
a	T	T
c	T	
b		T
g		T

(b)

The *ITL-list* as shown in Fig. 3a (which is the result after processing the second transaction) contains *ItemN*, *CountN*, *RangeC*, *FirstT*, *LastT*, *Mark*, and *BP* for each pattern. It is used to record the useful information of each pattern. Take Table 2 as an example. Assume that we have the *MinSup* = 3, the *MinPR* = 75%, and the *MaxPer* = 2. First, we scan on the first transaction, "1: *ac*" with *ts* = 1. Since pattern *a* and pattern *c* are the first occurrence, we create their own *ITL-list* because the *ts* (timestamp) of the first occurrence is not larger than the *MaxPer* and the *RangeC* plus 1. Therefore, we modify their *CountN*, *RangeC*, *FirstT*, and *LastT* to 1, 1, 1, and 1, respectively. Since *FirstT* only records the *ts* of the first occurrence, even if it appears later, we do not need to deal with this variable. Then, we have to deal with the *BP* part as shown in Fig. 3b, where *BP* is a table composed of a series of items and Boolean values. For the above example, we know that pattern *a* and pattern *c* appear, when *ts* is 1. So in their *BP* table, we will set the Boolean value to *true* at T1.

Next, we scan the second transaction, "2: *abg*" with *ts* = 2. Since pattern *b* and pattern *g* are the first occurrence, we create their own *ITL-list*, because *ts* of the first occurrence is equal to the *MaxPer* and the *RangeC* plus 1. Therefore, we modify their *CountN*, *RangeC*, *FirstT*, and *LastT* to 1, 1, 2, and 2, respectively. Then, the

ItemN	CountN	RangeC	FirstT	LastT
a	8	9	1	12
c	8	8	1	13
b	7	8	2	12
g	2	1	2	8
d	6	6	3	12
e	5	3	3	13
f	4	4	8	13

ItemN	T1	T2	T3	T4	T5	T6	T7	T8	T9	T10	T11	T12	T13	Mark
a	T	T		T		T	T		T		T	T		T
c	T			T		T		T		T	T	T	T	T
b		T		T		T	T		T		T	T		T
g		T						T						F
d			T	T		T		T		T		T		T
e			T				T			T	T		T	T
f								T		T	T		T	T

Fig. 4 The complete *ITL-list* of each pattern at level 1

CountN, *RangeC*, and *LastT* values of pattern *a* are updated to 2, 2, and 2, respectively. Because the current *ts* minus the *LastT* of pattern *a* is less than *MaxPer*, *RangeC* is increased by 1. At the same time, for pattern *a*, we set T2 in *BP* to *true*, and for patterns *b* and *g*, we do the same thing. The result is shown in Fig. 3a, b.

The same step is processed until the entire database is scanned, and the variables which have not been mentioned, Mark, will not be decided until the whole database has been scanned. For instance, through Table 2, we can know that pattern *a* last appears in ts 12. If the difference between the timestamp of the last transaction of the database and *LastT* of the pattern is less than or equal to *MaxPer*, we set Mark to True and increase *RangeC* by 1. Take pattern *a* as an example. The gap between 13 (ts of the last transaction) and 12 (*LastT* of "a") is less than *MaxPer*, so Mark of "a" is set to True and we increase its *RangeC* by 1. The complete *ITL-list* of each pattern at level 1 is shown in Fig. 4.

Let's start with a transactional database TDB3 to construct the *ITL-tree* as shown in Fig. 5 (which is the result after processing the second transaction) and the *ITL-list*. Basically, there are two cases which we must concern: (1) the item never appears before; (2) the item has appeared before. For the first case, the item never appears before; we add a new node at level 1 of the *ITL-tree* and give it a name. At the same time for this node, we create its *ITL-list*. For the second case, the item has appeared before, we need to find the node with this name from level 1 in the *ITL-tree*, and then update its *ITL-list*. When we scan the first transaction, we can know that there are

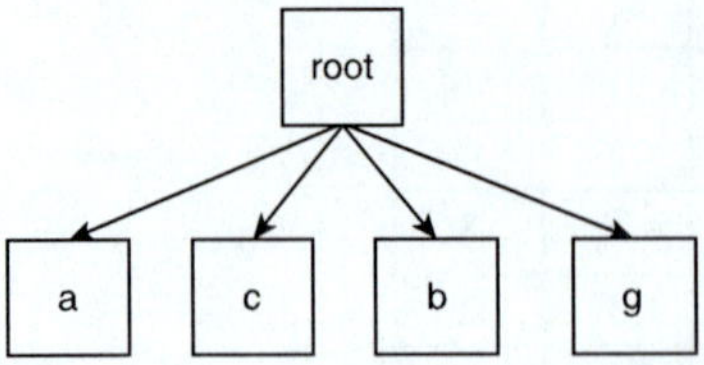

ItemN	CountN	RangeC	FirstT	LastT	T1	T2
a	2	2	1	2	T	T

ItemN	CountN	RangeC	FirstT	LastT	T1
c	1	1	1	1	T

ItemN	CountN	RangeC	FirstT	LastT	T1	T2
b	1	1	2	2		T

ItemN	CountN	RangeC	FirstT	LastT	T1	T2
g	1	1	2	2		T

Fig. 5 The *ITL-tree* and the *ITL-list* after scanning the second transaction

two patterns in the transaction, which are patterns *a* and *c*. When the algorithm reads the items in a transaction, it will change the *ITL-list* of these patterns in the *ITL-tree*. For instance, pattern *a* is the first pattern to be scanned, so we insert node *a* into level 1 of the *ITL-tree*. At the same time, we will create an *ITL-list* belonging to pattern *a* and start to modify the data inside. The *CountN* is set to 1, because pattern *a* appears for the first time. The *RangeC* is set to 1, because its period (i.e., 1–0 = 1, where 0 is the starting time) is less than *MaxPer*. The *FirstT* is set to 1, because the first occurrence is at *ts* 1. The *LastT* is set to 1, because the last occurrence is at *ts* 1. Next, we set the T1 of *BP* table to True. The remaining items in the first transaction is executed by the same way.

Next, we scan the second transaction. The second transaction has patterns *a*, *b*, and *g*. Since pattern *a* already exists at level 1 of the *ITL-tree*, we only need to update the data of its *ITL-list*. Take pattern *a* as an example. In the *ITL-list* of pattern *a*, its *CountN* is set to 2. Because it has appeared once before, we have 1 + 1 = 2. The *RangeC* is set to 2, because its period (i.e., 2–1 = 1, where the first 1 is *LastT*) is less than *MaxPer*. The *LastT* is set to 2, because the pattern *a* appears at *ts* 2 now. Then, for T2 of *BP* table, we set its *BP* table entry to True. Next, we turn to *b* and *g*. Take pattern *b* as an example. Since pattern *b* appears for the first time, we insert node *b* into level 1 of the *ITL-tree* and create its *ITL-list*. Then, we set its *CountN* to 1. The *RangeC* is set to 1, because its period (i.e., 2–0 = 2, where 0 is the starting time) is equal to *MaxPer*. The *FirstT* is set to 2, because the first occurrence is at *ts* 2. The *LastT* is set to 2, because the last occurrence is at *ts* 2. Moreover, we set the T2 of *BP* table to True. Since "b" does not appear at *ts* 1, T1 of *b* is empty. The condition of pattern *g* is the same as pattern *b*, so we do what has done for pattern *b* once for pattern *g*. As a result, the *ITL-tree* and the *ITL-list* after scanning the second transaction is shown in Fig. 5.

Finally, we perform the above process until the entire database has been scanned to complete level 1 of the *ITL-tree*, and the *ITL-list* are shown in Fig. 6. But we have not gotten the last period from the last *ts* (13). This last period will determine whether we have to increase *RangeC* by 1 and set *Mark* to *True*. At the same time, we can also know which of the patterns at level 1 are noncandidate patterns, candidate patterns, and partial periodic-frequent patterns.

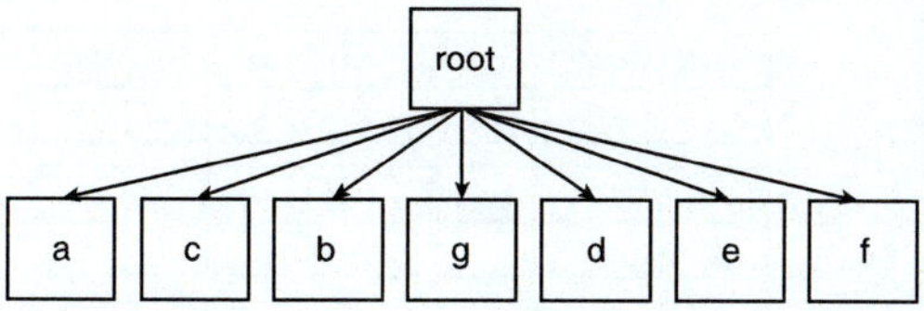

ItemN	CountN	RangeC	FirstT	LastT	BP
a	8	8	1	12	...

ItemN	CountN	RangeC	FirstT	LastT	BP
c	8	7	1	13	...

ItemN	CountN	RangeC	FirstT	LastT	BP
b	7	7	1	12	...

ItemN	CountN	RangeC	FirstT	LastT	BP
g	2	1	2	8	...

ItemN	CountN	RangeC	FirstT	LastT	BP
d	6	5	3	12	...

ItemN	CountN	RangeC	FirstT	LastT	BP
e	5	2	3	13	...

ItemN	CountN	RangeC	FirstT	LastT	BP
f	4	3	8	13	...

Fig. 6 The *ITL-tree* and the *ITL-list* after scanning the entire database

Therefore, the modified *ITL-tree* and the modified *ITL-list* are shown in Fig. 7. Through Fig. 7, according to the *ITL-list* of each node, we can know, whether this node can become the partial periodic-frequent pattern. As long as the number of *RangeC* in the *ITL-list* is not less than *MinPR* × (*MinSup* + 1) and *CountN* is not less than *MinSup*, we set it to the dotted circle. Furthermore, if *RangeC* is not less than *MinPR* × (*CountN* + 1), we set it to the solid circle, and this pattern is the partial periodic-frequent pattern which we need. The remaining square boxes are considered unnecessary, because its *CountN* is less than *MinSup* or its *RangeC* is less than *MinPR* × (*MinSup* + 1). So even if it is merged with other patterns, it cannot be a partial periodic-frequent pattern. Therefore, the partial periodic-frequent size 1 patterns are patterns *a*, *c*, *b*, *d*, and *f*.

Since we have found all partial periodic-frequent size 1 patterns, we can start finding partial periodic-frequent patterns of size 2. First, we will exploit these dotted or solid circle patterns in Fig. 7, where *RangeC* is larger than or equal to the *MinPR* × (*MinSup* + 1) and *CountN* is not less than *MinSup*. We merge two patterns on the same level of the *ITL-tree* with the same prefix into a (level + 1) super-set patterns. Note that the prefix can be null and we do not need to merge with the square box, because it is impossible to generate the partial periodic-frequent pattern. Since we want to merge the two patterns, we need to decide the order of the merged patterns. In the process of generating the size 2 patterns, we will merge the patterns at level 1 from left to right. Take Fig. 7 as an example. Patterns at level 1 of the *ITL-tree* have the order from left to right: [*a*, *c*, *b*, *d*, *e*, *f*]. Moreover, the pattern will generate a new pattern with each pattern after its order, and then it will be the next pattern. For instance, pattern *a* will generate the new size 2 pattern with pattern *c*, *b*, *d*, *e*, *f*, and then it will turn to pattern *c* and pattern *b*, *d*, *e*, *f* to generate the new size 2 pattern, and so on.

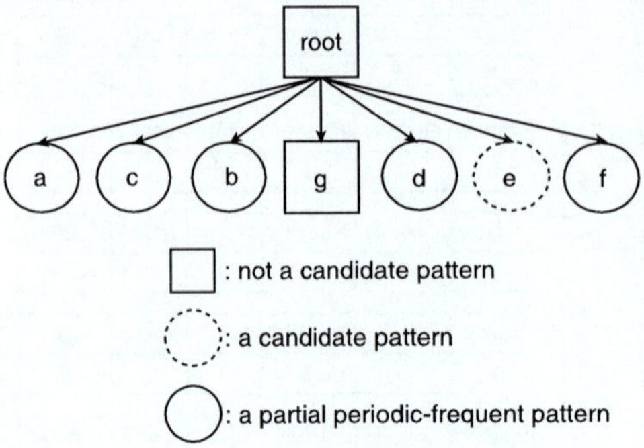

ItemN	CountN	RangeC	FirstT	LastT	BP	Mark
a	8	9	1	12	...	T

ItemN	CountN	RangeC	FirstT	LastT	BP	Mark
c	8	8	1	13	...	T

ItemN	CountN	RangeC	FirstT	LastT	BP	Mark
b	7	8	2	12	...	T

ItemN	CountN	RangeC	FirstT	LastT	BP	Mark
g	2	1	2	8	...	F

ItemN	CountN	RangeC	FirstT	LastT	BP	Mark
d	6	6	3	12	...	T

ItemN	CountN	RangeC	FirstT	LastT	BP	Mark
e	5	3	3	13	...	T

ItemN	CountN	RangeC	FirstT	LastT	BP	Mark
f	4	4	8	13	...	T

Fig. 7 The modified *ITL-tree* and the modified *ITL-list*

Begin (at T8) End (at T12)

	T1	T2	T3	T4	T5	T6	T7	T8	T9	T10	T11	T12	T13
a	T	T		T		T	T		T		T	T	
f								T		T	T		T

Fig. 8 The *BP* table for patterns *a* and *f*

Next, we illustrate how to generate size 2 patterns at level 2 of the *ITL-tree* and complete their *ITL-list*. Take pattern *a* and *f* as an example. At first, since we cannot find the pattern *af* connected to *a*, we create a node *af* and connect it to *a* with an edge. Then, we compare the *FirstT* of patterns *a* and *f* to find the larger value. This value means that when we check the *BP* table of two patterns, we only need to check from the Boolean value at this position. Because it will never appear at the same timestamp before. Similarly, we compare the *LastT* of the patterns *a* and *f* to find the smaller value. This value means that when we check the *BP* table of two patterns, we only need to check the Boolean value at this position from the front. Because after that, they will never appear at the same timestamp. Take Fig. 8 as an example. From this figure, we can know that pattern *f* does not appear before *ts* 8 and *a* does not appear after *ts* 12. Therefore, we only need to check the Boolean value between *ts* 8 and *ts* 12, which avoids unnecessary actions.

Then, we do the *AND* operation on *BPs* of patterns *a* and *f*. This *BP* table after the *AND* operation is regarded as *BP* table of pattern *af*. We will use this *BP* table to complete its *ITL-list*. At first, we need to find the timestamp with a *True* value

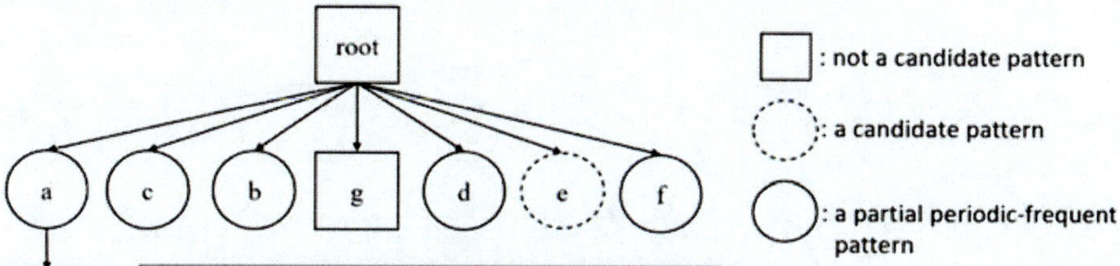

ItemN	CountN	RangeC	FirstT	LastT	BP	Mark
af	1	1	11	11	...	T

Fig. 9 The *ITL-tree* and the *ITL-list* after the pattern *af* is inserted

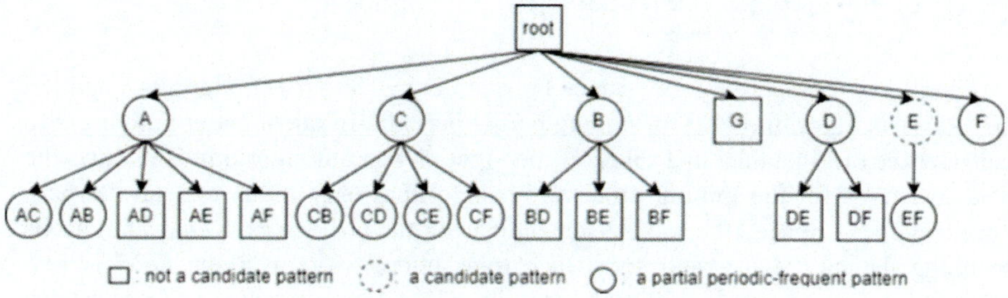

Fig. 10 The size 2 patterns of the *ITL-tree*

between *ts* 8 and *ts* 12. Then, we know that the first *ts* with *True* is at *ts* 11. Therefore, we set its *CountN*, *RangeC*, *FirstT*, and *LastT* to 1, 0, 11, and 11, respectively. Then, we find that the following timestamp has no *True*. Therefore, as long as we find the last period to modify the value of *RangeC* and *Mark*, we complete the *ITL-list* of *af*. At the same time, we can confirm whether pattern *af* is the noncandidate pattern, the candidate pattern, or the partial periodic-frequent pattern through the *ITL-list*, which is shown in Fig. 9. We apply the above steps to the remaining patterns. The result is shown in Fig. 10. The partial periodic-frequent size 2 patterns are patterns *ac*, *ab*, *cb*, *cd*, *ce*, cf, and *ef*.

After mining all the partial periodic-frequent size 2 patterns, we start to find the partial periodic-frequent size 3 patterns. Similarly, we will merge the two patterns with the same prefix at level 2 of the *ITL-tree* except for the noncandidate patterns. So based on the above sentence, we will insert the new size 3 patterns *acb*, *cbd*, *cbe*, *cbf*, *cde*, *cbf*, and *cef* into level 3 of the *ITL-tree* and follow the previous steps to complete their *ITL-list*.

Then, we confirm the type of each pattern according to their *ITL-list*. The result is shown in Fig. 11. According to this figure, we can know that patterns *acb* and *cef* are partial periodic-frequent patterns. Moreover, there is no size 4 pattern which can be generated. So all the partial periodic-frequent patterns in this database have been found.

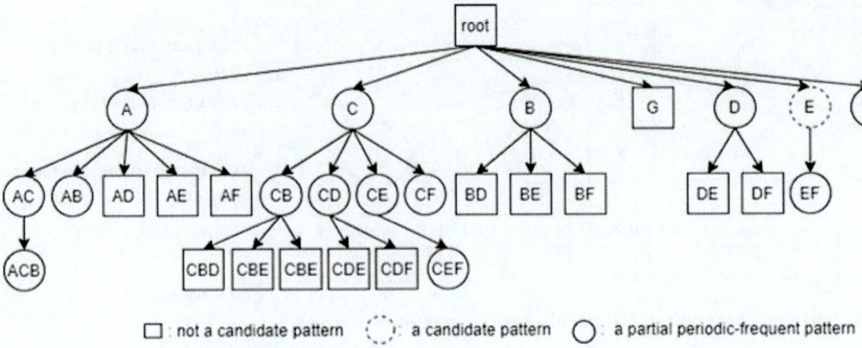

Fig. 11 The size 3 patterns of the *ITL-tree*

Finally, we compare the difference between our *Inverted-ITL* algorithm and the *GPF-growth* algorithm [12] for mining partial periodic-frequent pattern mining. We consider the original data in Table 2 as the input. When our algorithm constructs the data structures for the mining process, our algorithm only needs to scan database once and does not need the sorting step. However, the *GPF-growth* algorithm needs scanning database twice and sorts each transaction. Moreover, the *GPF-growth* algorithm needs to generate a prefix tree based on the reordered database, and it will generate many prefix trees during the mining process.

4 Performance

4.1 The Performance Model

We compare the processing time of the *Inverted-ITL* algorithm and the *GPF-growth* algorithm [12]. for mining partial periodic-frequent patterns in the real and synthetic datasets. We will consider using different values of the minimum support threshold, the maximum period threshold, and the minimum periodic-ratio threshold to execute different size datasets on the algorithm. For the real datasets, we use the Retail dataset (http://fimi.ua.ac.be/data/) for experiments. The details of the dataset Retail contains the transaction count = 88,162, the item count = 16,470, and the average item count per transaction = 10.30. For the synthetic dataset, T10.I4.D100K was generated by using the generator from the IBM Quest Dataset Generator. The parameter *T*, *I*, and *D* represents the average item count per transaction, the average maximal size of frequent itemsets, and the number of transactions in the dataset, respectively.

4.2 Experiments Results

First, let's deal with the Retail dataset. In Fig. 12, we show the comparison of the processing time for the Retail dataset and the synthetic dataset under the change of the maximum period threshold. In this experiment, we set the minimum support threshold = 0.01% and the minimum periodic-ratio threshold = 0.01%. Through Fig. 12a, we observe that both our algorithm and the *GPF-growth* algorithm maintain a fairly stable curve. The reason is that all items in the Retail dataset, except for the few specific items, the rest of the items will basically not repeat within the threshold, so this situation will happen. Moreover, according to this figure, it shows that we provide better performance than the *GPF-growth* algorithm which scans the dataset twice and sorts many times. From Fig. 12b, we observe that the performance of our algorithm is also better than that of the *GPF-growth* algorithm.

In Fig. 13, we show the comparison of the processing time for the Retail dataset and the synthetic dataset under the change of the minimum support threshold. In this experiment, we set the maximum period threshold = 10 and the minimum periodic-ratio threshold = 0.01%. Through Fig. 13a, we observe that as the value of minimum support threshold increases, the processing time of the two algorithms decreases. The reason is that only a small number of patterns can meet the threshold of support, when the minimum support threshold increases. When the number of the candidate patterns continues to decrease, the processing time also decreases. Through this figure, we observe that when the number of candidate patterns continues to decrease, the time curve of the *GPF-growth* algorithm will change greatly. However, our algorithm changes relatively small. This is because as long as the number of candidate patterns increases, the number of times that the *GPF-growth* algorithm needs to be sorted and the number of prefix trees generated will also increase. However, our algorithm only needs to process the *ITL-list* of each candidate pattern which are already in the *ITL-tree*. Therefore, in Fig. 13a, the processing time of our algorithm is faster than the *GPF-growth* algorithm. In Fig. 13b, the performance of our algorithm is also better than that of the *GPF-growth* algorithm.

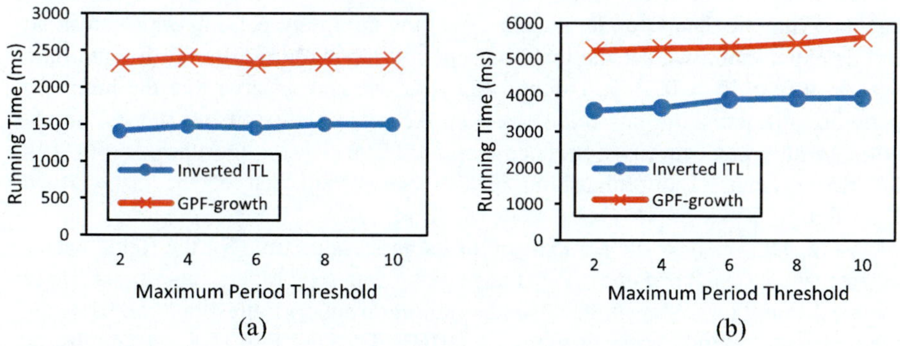

Fig. 12 A comparison of the processing time under the change of the maximum period threshold: (**a**) for the Retail dataset; (**b**) for the synthetic dataset

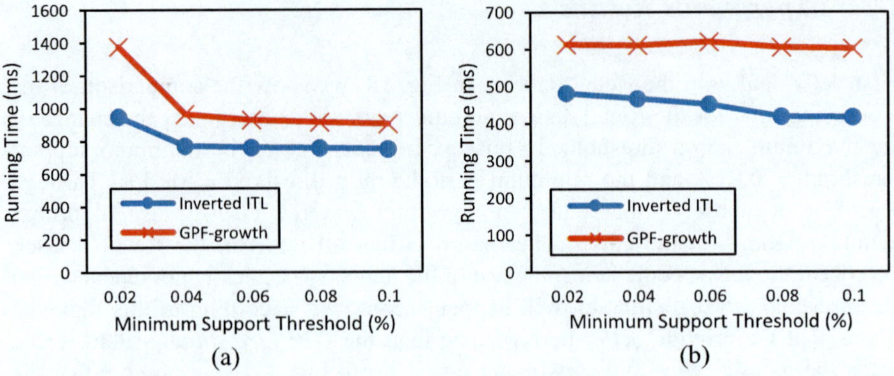

Fig. 13 A comparison of the processing time under the change of the minimum support threshold: (**a**) for the Retail dataset; (**b**) for the synthetic dataset

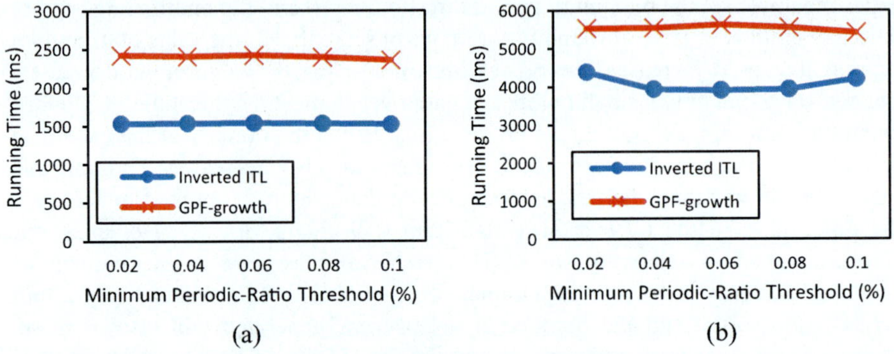

Fig. 14 A comparison of the processing time under the change of the minimum periodic-ratio: (**a**) for the Retail dataset; (**b**) for the synthetic dataset

In Fig. 14, we show the comparison of the processing time for the Retail dataset and the synthetic dataset under the change of the minimum periodic-ratio threshold. In this experiment, we set the maximum period threshold = 10 and the minimum support threshold = 0.01%. Through Fig. 14a, we can observe that the minimum periodic-ratio threshold has small effect on the performance of two algorithms. So our algorithm has better performance than the *GPF-growth* algorithm. In Fig. 14b, the performance of our algorithm is also better than that of the *GPF-growth* algorithm.

In Fig. 15, we show the comparison of the processing time for the Retail dataset and the synthetic dataset under the change of the data size. In this experiment, we set the maximum period threshold = 10, the minimum support threshold = 0.01%, and the minimum periodic-ratio threshold = 0.01%. Through Fig. 15a, we can observe

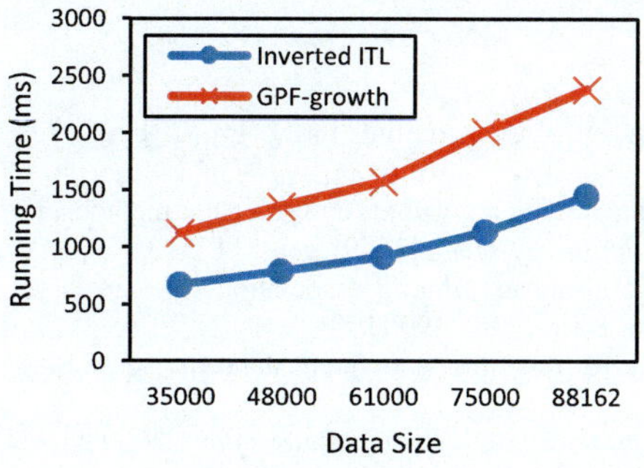

(a)

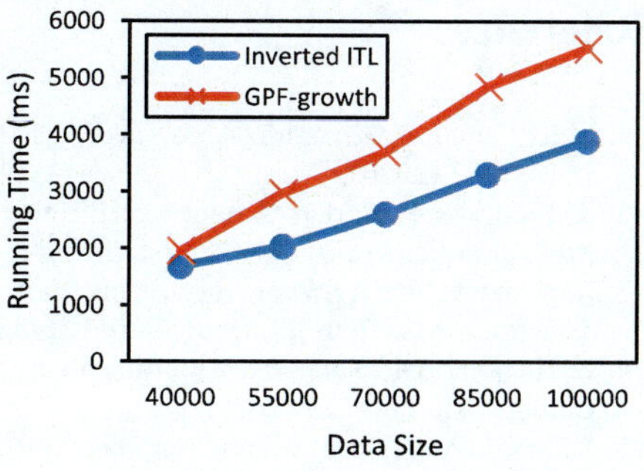

(b)

Fig. 15 A comparison of the processing time under the change of the change of the data size: (**a**) for the Retail dataset; (**b**) for the synthetic dataset

that as the size of the dataset increases, the processing time also increases. However, the growth rate of the *GPF-growth* algorithm is significantly larger than that of our algorithm. Because as the size of the dataset increases, the *GPF-growth* algorithm needs to sort the items in more transactions, but our algorithm only needs to deal with the *ITL-list* of each candidate pattern in the *ITL-tree*. Therefore, the *GPF-growth* algorithm needs a lot of time to deal with sorting and generating prefix trees, and we only need to deal with each pattern in the *ITL-tree*. Through Fig. 15b, the performance of our algorithm is also better than that of the *GPF-growth* algorithm.

5 Conclusion

In this paper, we have proposed an *Inverted-ITL* algorithm which can efficiently mine the partial periodic-frequent patterns. In the data mining, we have constructed the *ITL-tree* and the *ITL-list* which need less processing time to store information of each pattern than the *GPF-growth* algorithm for the same transactional database. From our simulation result, we have shown that our algorithm is more efficient than the *GPF-growth* algorithm.

Acknowledgments This research was supported in part by the National Science Council of Republic of China under Grant No. MOST 110-2221-E-110-054.

References

1. M.K. Gupta, P. Chandra, A comprehensive survey of data mining. Int. J. Inf. Technol. **12**, 1243–1257 (2020)
2. K. Kaithwas, P. Borkar, A review on different data mining algorithms and selection methods, in *Proc. of the 2019 Int. Conf. on Intelligent Sustainable Systems*, (2019), pp. 511–515
3. R. Agrawal, R. Agrawal, R. Srikant, Fast algorithms for mining association rules in large databases. Proc. 20th Int. Conf. Very Large Data Bases, 487–499 (1994)
4. Z.-H. Deng, DiffNodesets: An efficient structure for fast mining frequent itemsets. Appl. Soft Comput. **41**, 214–223 (2016)
5. J. Han, J. Pei, Y. Yin, Mining frequent patterns without candidate generation. Proc. 2000 ACM SIGMOD Int. Conf. Manag Data **29**(2), 1–12 (2000)
6. H. Bui, B. Vo, H. Nguyen, T.-A. Nguyen-Hoang, T.-P. Hong, A weighted N-list-based method for mining frequent weighted itemsets. Expert Syst. Appl. **96**, 388–405 (2018)
7. T. Le, B. Vo, An N-list-based algorithm for mining frequent closed patterns. Expert Syst. Appl. **42**(19), 6648–6657 (2015)
8. B. Vo, S. Pham, T. Le, Z.-H. Deng, A novel approach for mining maximal frequent patterns. Expert Syst. Appl. **73**, 178–186 (2017)
9. R.U. Kiran, M. Kitsuregawa, Novel techniques to reduce search space in periodic-frequent pattern mining. Proc. 19th Int. Conf. Database Syst. Adv. Appl., 377–391 (2014)
10. R.U. Kiran, M. Kitsuregawa, P.K. Reddy, Efficient discovery of periodic-frequent patterns in very large database. J. Syst. Software **112**, 110–121 (2016)
11. R.U. Kiran, P.K. Reddy, Towards efficient mining of periodic-frequent patterns in transactional databases. Proc. 21th Int. Conf. Database Exp Syst Appl, 194–208 (2010)
12. R.U. Kiran, J.N. Venkatesh, M. Toyoda, M. Kitsuregawa, P.K. Reddy, Discovering partial periodic-frequent patterns in a transactional database. J. Syst. Software **125**, 170–182 (2017)
13. A. Surana, R.U. Kiran, P.K. Reddy, An efficient approach to mine periodic- frequent patterns in transactional databases. Proc. 15th Pacific-Asia Conf. Knowl. Discov. Data Mining, 254–266 (2011)
14. S.-S. Chen, T.C.-K. Huang, Z.-M. Lin, New and efficient knowledge discovery of partial periodic patterns with multiple minimum supports. J. Syst. Software **84**(10), 1638–1651 (2011)

A Viewpoints-Based Analysis of Enterprise Architecture Debt

Ather Maqsood, Peter Alexander, Horst Lichter, and Sansiri Tanachutiwat

Abstract Technical debt solely focuses on the technical aspect and fails to provide a holistic view to address the misalignment between business and IT aspects of enterprise architecture (EA). To provide a more holistic view and to include the business aspect in the context of debt, researchers have proposed the concept of EA debt. It refers to the gap between the present state and the hypothetically ideal state of EA. EA debt, if accumulated, has negative impacts on the enterprise and can result in a decrease in the value, maintainability, agility, and efficiency of EA. Therefore, it is very crucial to manage EA debt. Furthermore, each stakeholder in the enterprise has an interest in certain aspects and areas of EA. By applying viewpoints, stakeholders can analyze EA debt by focusing on aspects and areas of their interest. By analyzing EA debt, stakeholders can make informed decisions during EA debt management (EADM). This paper presents a systematic mapping study carried out to collect predefined viewpoints from areas of enterprise architecture, software architecture, and technical debt. The collected viewpoints are then assessed to determine if they can be used for the analysis of EA debt. Further, six categories are defined to categorize EA debt viewpoints based on their purpose and content.

Keywords Enterprise architecture debt · Enterprise architecture debt management · Enterprise architecture management · Viewpoints

---

A. Maqsood (✉) · S. Tanachutiwat
The Sirindhorn International Thai-German Graduate School of Engineering (TGGS), KMUTNB North Bangkok, Bangkok, Thailand
e-mail: ather.m@email.kmutnb.ac.th; sansiri.t@tggs.kmutnb.ac.th

P. Alexander · H. Lichter
Research Group Software Construction RWTH Aachen University, Aachen, Germany
e-mail: peter.alexander@swc.rwth-aachen.de; lichter@swc.rwth-aachen.de

C.-C. Wang, A. Nallanathan (eds.), *Proceedings of the 5th International Conference on Signal Processing and Information Communications*, Signals and Communication Technology, https://doi.org/10.1007/978-3-031-13181-3_11

1 Introduction

Enterprise architecture (EA) is a discipline that focuses on improving the alignment between the business and IT aspects of enterprise [1]. IT supports an enterprise's business processes; thus, for success, it must be aligned with the business of the enterprise [2]. The process of developing and maintaining a coherent enterprise is complicated as it involves various stakeholders, each having a different understanding and perspective about the architecture [3]. Furthermore, factors such as stakeholder expectations, technological advancements, and organizational strategies or goals impact the alignment process [2]. Consequently, the misalignment between business and IT becomes frequent.

The concept of technical debt focuses on the technical aspect in the misalignment between business and IT. This concept was introduced by Cunningham [4] to address the repercussions of weak software development. Fernandez-Sanchez et al. [5] discussed that technical debt refers to the consequences of sub-optimal software development and it is used to make extra costs visible, which occur due to the poor quality of software. Furthermore, if accumulated, technical debt results in negative consequences such as an increase in software development cost, decline in maintainability, decrease in product quality, and slow progress of the software development [6].

Technical debt has a limitation that it solely focuses on the technical aspect of EA and does not address the business aspect [7]. To overcome this limitation, and to provide a holistic view in addressing the misalignment between IT and business, Hacks et al. [7] proposed the concept of EA debt, which refers to the gap between the present state and hypothetically ideal state of EA [7].

EA debt, if accumulated, has adverse effects on the business of an enterprise. Yeong et al. [8] discussed that managing EA debt is one of the success factors for an effective EA implementation because it increases awareness and gives insight into sub-optimal implementations in the architecture. Therefore, it should be managed by allocating resources [8]. Furthermore, EA debt has many aspects, and it could arise from any part of the architecture from any layer of EA [7]. As each stakeholder has an interest in specific aspects and areas of architecture, an approach is required using which stakeholders can gain insight into EA debt by focusing on certain aspects and areas of EA.

Viewpoints address particular concerns of stakeholders, and each viewpoint focuses on a specific aspect of architecture [3]. By applying viewpoints stakeholders can gain insight into divergent solutions by focusing on specific aspects and areas of interest. However, since the concept of EA debt is relatively new, there are no viewpoints defined explicitly for this specific area. Therefore, a systematic mapping study is carried out to collect viewpoints from areas closer to EA debt.

This study will answer the following research question and pertinent sub-research questions:

RQ-1: *What are the viewpoints that could be used in the context of EA debt analysis?*
RQ-1.1: *What are the existing viewpoints in the areas closely related to EA debt?*
RQ-1.2: *How to derive viewpoints that are relevant for analyzing EA debt from some existing viewpoints used in related areas of study?*
RQ-1.3: *How to document a viewpoint of EA debt?*

The rest of the paper is organized as follows: Section 2 describes the background of the study. Section 3 talks about the research methodology, and Sect. 4 discusses the results of the study. Furthermore, Section 5 focuses on the evaluation of the idea of viewpoint-based analysis of EA debt. Section 6 discusses how the study results answer the research question. Also, it discusses the implications of results and the threats to validity. Lastly, Section 7 talks about the conclusion and future work.

2 Background

The concept of technical debt was first introduced by Cunningham [4] to refer to the consequences of weak software development and the trade-offs between technical development activities that are delayed to get short-term payoffs [9]. It has detrimental effects such as increased cost, low product quality, and slow progress to the long-term success of software development [10].

Technical debt focuses solely on the technical aspect; hence, it fails to consider the business aspect while addressing business and IT misalignment [7]. Therefore, to provide a holistic view in the context of debt, Hacks et al. [7] proposed the concept of EA debt and defined it as "a metric that depicts the deviation of the currently present state of an enterprise from a hypothetical ideal state."

EA debt arises when debt is taken in an artifact. This implies that the artifact is not implemented or executed optimally as compared to an ideal situation [7]. Furthermore, to have a clear overview of debt, each artifact and part of the artifact should be weighted as per an organization's defined weighting methodology [7]. Moreover, EA debt, if accumulated, can have negative effects on the enterprise [7]. Consequently, the enterprise would have to put efforts and resources into managing it. Therefore, it is imperative to manage EA debt.

Alexander et al. [11] proposed a framework for managing EA debt. It consists of nine main activities and many of them are embraced from the technical debt management activities identified in [12]. The activities are identification of EA debt, collection of evidence for suspicions of EA debt, assessment of consequences of the identified EA debts, prioritization EA debts, monitoring of the changes in the EA, repayment or prevention activity to mitigate effects of EA debt, and lastly, documentation and communication of EA debt information [11].

The complex EA debt information can be presented in various views by applying viewpoints. Viewpoints frame specific concerns of stakeholders and allow them to view certain aspects in isolation, as well as compare more than one aspect [3]. Further, viewpoints are communicative in nature and used for various purposes such as design, analysis, communication, and decision-making [13]. Moreover, the architectural information obtained by looking from a viewpoint is referred to as view, and the point from which architecture is investigated is referred to as viewpoint [3].

The concept of viewpoint is used in various domains for various purposes. For example, in the domain of software architecture, Bedjeti et al. [14] proposed a viewpoint to describe the context of software systems. Further, in the domain of information systems, Närman et al. [15] proposed a set of viewpoints for the analysis of information systems. Furthermore, in the domain of EA, Miguens et al. [16] proposed a viewpoint to represent costs, P. Clements and L. Bass [17] proposed a viewpoint to describe business goals, and Da Silva et al. [18] proposed a viewpoint for analysis of EA evolution. Also, in the domain of technical debt, Li et al. [19] proposed a set of viewpoints to document architectural technical debt.

In the context of EADM, an approach is needed using which stakeholders can analyze EA debt. By applying viewpoints, stakeholders can gain insight into EA debt by concentrating on specific aspects and areas of the architecture. This would enable stakeholders to obtain information about EA debt that is of relevance to them. As a result, stakeholders can analyze EA debt and make informed decisions to manage EA debt based on information obtained by applying viewpoints.

This research contributes to the area of EA debt by presenting viewpoints from the literature which can be applied for the analysis of EA debt. Furthermore, the applicable viewpoints are documented online in [20]. This online repository may serve as the knowledge base for EA debt viewpoints.

The concept of EA debt is relatively new, and to our knowledge, no approach has been provided in the literature that would assist stakeholders in the analysis of EA debt by allowing them to focus on aspects and areas of architecture that are of relevance to them.

Yeong et al. [8] proposed a portfolio-based approach to prioritize EA debt. Further, this approach assists stakeholders in reasoning their EA investment decisions, and it encompasses portfolio thinking and utility theory [8]. Moreover, in the domain of technical debt, to assist stakeholders in understanding architectural technical debt (ATD), Li et al. [19] proposed a set of architecture viewpoints that frame concerns related to ATD. By employing the ATD viewpoints, the impact of ATD can be communicated in decision-making, and the knowledge of ATD can be made available to relevant stakeholders [19]. Further, to support stakeholders in the decision-making about technical debt, Reboucas de Almeida [21] proposed a business-driven approach. This approach is based on a decision-making framework that prioritizes technical debt based on how IT assets support organizations' business processes [21]. Furthermore, Martini and Bosch [22] proposed a method named as AnaConDebt that helps stakeholders to analyze the impact of ATD. The method provides indicators that estimate crucial aspects responsible for interest growth and alerts stakeholders if refactoring is required [22]. Also, to assist stakeholders in understanding the causes and effects of technical debt, Rios et al. [23] proposed employing probabilistic cause-effect diagrams for the cause-effect analysis of technical debt. These diagrams illustrate the causes that may have contributed most to the occurrence of the analyzed problem as well as the most prevalent effects that arise as a result of the problem [23]. As well, Zazworka et al. [9] proposed a methodology for cost-benefit analysis of technical debt that is based on metric analysis and software repository mining. It aids in the identification of refactoring activities based on their expense and potential impact [9].

3 Methodology

The research methodology used in this study is the design science research methodology (DSRM) [24]. The goal of this research methodology is to develop solutions that are intended to solve identified problems, and it consists of six activities. Further, the activities proposed in DSRM, which are employed in this work, are the following:

3.1 Problem Identification and Motivation

EA debt, if accumulated, has adverse effects on the business of an enterprise. Further, EA debt has multiple aspects, and it may originate from any part of the EA. As well, there are various stakeholders in an enterprise, and every stakeholder has different knowledge, objectives, responsibilities, and areas of interest in the architecture. Therefore, in EADM, an approach is required using which stakeholders can analyze EA debt by focusing on certain aspects and areas of architecture.

3.2 Objectives Definition

The objective of the solution is to offer an approach that would enable stakeholders to analyze EA debt by focusing on specific aspects and areas of the architecture that are of relevance to them. Furthermore, stakeholders should be able to filter out information about EA debt that is irrelevant to them. As a result, they would be enabled to make informed decisions during the EADM process.

3.3 Design and Development

The solution artifact suggested is the idea of viewpoint-based analysis of EA debt. By applying viewpoints, the complex EA debt information can be structured in various views addressing deviations in solutions by focusing on various aspects and areas of architecture. This way, the complex EA debt information can be split into various views, and stakeholders can acquire architectural information about EA debt that is of interest to them.

Since the concept of EA debt is still relatively new, there are no viewpoints explicitly defined for this area. Therefore, a thorough mapping study is conducted to collect viewpoints from areas closely associated with EA debt. After this, the viewpoints collected are filtered for their applicability for EA debt analysis. Subsequently, the applicable viewpoints are documented. Moreover, these steps have been discussed in the following subsections:

Systematic Mapping Study Systematic mapping study is described as a form of secondary study to overview a particular topic, find research gaps, and gather evidence that would be used for future research [25]. Generally, systematic mapping studies have a broader research question and data extraction process [25].

The study is executed by following the guidelines proposed for systematic mapping study by Kitchenham and Charters [26] and Petersen et al. [25]. Furthermore, the steps in study execution are as follows:

Research Question To achieve the goal of collecting *viewpoints* from literature, that can be applied for EA debt analysis, the mapping study will answer the following sub-research question:

- *RQ-1.1: What are the existing viewpoints in the areas closely related to EA debt?*

Study Search The study's goal was to collect viewpoints from enterprise architecture, software architecture, and technical debt. Therefore, these terms are included as keywords in the search query. Further, for technical debt, similar to the mapping study in [12], the term "debt" is used in the search query. This is because technical debt has many types, and in primary studies, it may be referred to by the name of its types. Furthermore, the final search query employed in electronic databases is as follows: search query ("software architecture" or "enterprise architecture" or "debt") and "viewpoint").

The primary studies for systematic mapping study were decided to be searched in literature by employing automatic search in nine popular electronic databases as shown in Table 1. However, a trail search was conducted in each database to check the relevance of primary studies returned by the database. As a result of trail search, all databases except DB-9 were finalized for the search of primary studies. DB-9 was excluded because the primary studies returned by the database during the trial search were not relevant to the concept of viewpoints.

Table 1 List of electronic databases

#	Database	Selection
DB-1	Scopus	Yes
DB-2	IEEE Xplore	Yes
DB-3	Science Direct	Yes
DB-4	ACM Digital Library	Yes
DB-5	Springer Link	Yes
DB-6	Web of Science	Yes
DB-7	Dblp Link	Yes
DB-8	IET Digital Library	Yes
DB-9	Wiley Online	No

Table 2 Inclusion criteria for screening of papers

#	Inclusion criteria
I1	The study belongs to the area of software architecture, enterprise architecture, or technical debt
I2	The study is relevant to the concept of viewpoints
I3	The study proposes a viewpoint or set of viewpoints. Alternatively, it discusses pre-defined viewpoints

Table 3 Exclusion criteria for screening of papers

#	Exclusion criteria
E1	The study is not relevant to viewpoints
E2	The study proposes approaches based on the concept of viewpoints but does not discuss or propose specific viewpoints
E3	The study is in the form of abstracts, tutorials, or talks

Screening of Papers The primary studies that were gathered as the result of the automatic search in databases were filtered. To include relevant primary studies and exclude irrelevant primary studies for *the* final review, inclusion and exclusion criteria were formulated which are shown in Tables 2 and 3, respectively.

Similar to the mapping by Li et al. [12], the selection of primary studies was executed in three phases. During the first phase, the papers were filtered based on the meta-data, i.e., paper's title and keywords. As well, the inclusion criteria (I1, I2) and exclusion criteria (E1) were employed in this step. However, papers that were hard to decide by looking into meta-data were decided to be selected or rejected in the next selection phase. Further, in the second phase of selection, papers were selected or rejected by reading the abstracts. In this phase, inclusion criteria (I2, I3) and exclusion criteria (E2, E3) were employed. Lastly, in the third phase, papers were selected by reading the full text of the papers. During this phase, the inclusion criteria (I3) and exclusion criteria (E2) were employed. Furthermore, the overall studies selection process is shown in Fig. 1.

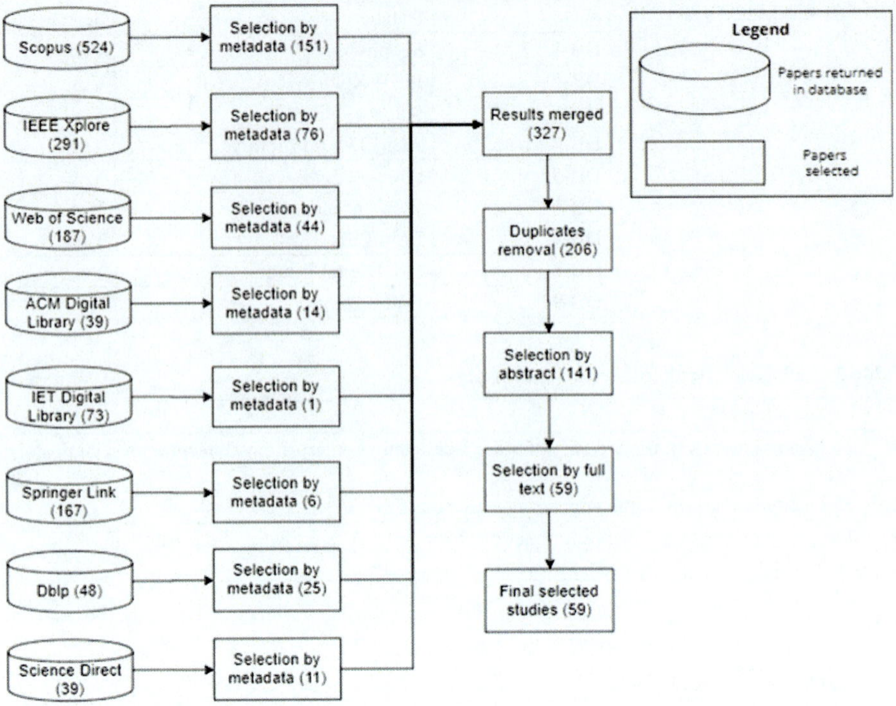

Fig. 1 Selection of primary studies

Data Extraction This step aims to define and extract the data items to be used to answer the research questions [26]. The data items to be extracted are first decided. After this, to reduce bias, a piloted study is conducted on some of the primary studies [26].

Moreover, to answer the research question, from each selected study, information of every entity that was used to document viewpoints (e.g., viewpoint description, concerns, stakeholders, meta-model, correspondence rules, operations on views) was extracted, and the data was recorded in an excel spreadsheet. However, in the majority of the primary studies, only the information about viewpoints description, concerns, and stakeholders was discussed.

Data synthesis. Data synthesis involves collecting, combining, and summarizing the results obtained from the data extraction step [26]. It aims to synthesize the data extracted to answer the research.

Moreover, frequency synthesis was used to identify the annual trend of publications and the distribution of the primary studies to the domains of enterprise architecture, software architecture, and technical debt. The annual trend of publication is shown in Fig. 2, and the distribution of studies to the different domains is shown in Fig. 3. Furthermore, the detailed results of the systematic mapping study are discussed in Sect. 4.

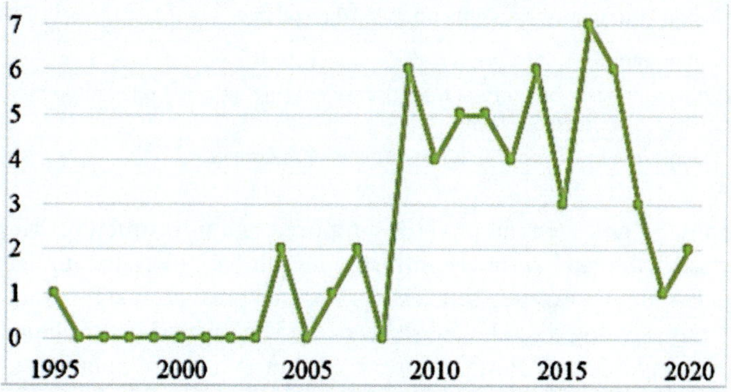

Fig. 2 Distribution of primary studies over the time

Fig. 3 Distribution of primary studies over domain types

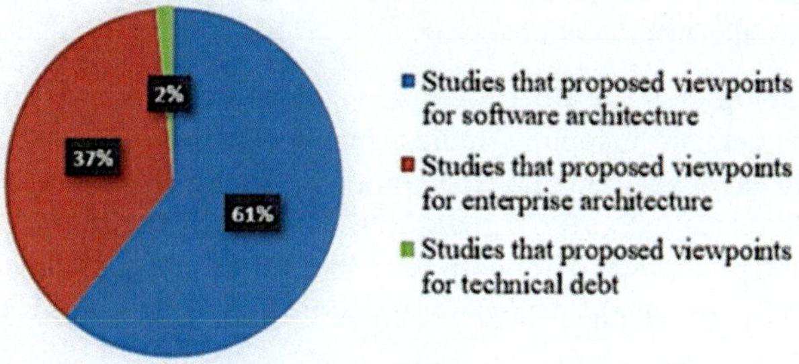

Table 4 Inclusion criteria of viewpoints for EA debt analysis

#	Inclusion criteria
IC-1	The viewpoint provides an understanding of divergent solutions by providing an insight into their design
IC-2	The viewpoint shows relations among divergent solutions across domain architectures and provides information that can assist in decision-making
IC-3	The viewpoint informs stakeholders by representing stakeholder relevant interpretations of divergent solutions
IC-4	The viewpoint can describe a divergent solution from a specific aspect or area of enterprise architecture
IC-5	The viewpoint can describe a divergent solution from a related set of aspects or areas of enterprise architecture
IC-6	The viewpoint can provide a holistic view of divergent solutions

Filtering of Viewpoints The viewpoints gathered by the systematic mapping study are then filtered for their applicability in the context of EA debt analysis. Further, for the filtering process, uniform inclusion and exclusion criteria are established. The inclusion criteria, as shown in Table 4, consist of factors that make a viewpoint applicable, and the exclusion criteria as shown in Table 5 consist of factors that make a viewpoint inapplicable for EA debt analysis.

Table 5 Exclusion criteria of viewpoints for EA debt analysis

#	Exclusion criteria
EC-1	The viewpoint describes software-level information that is not relevant to the concerns at the enterprise architecture level
EC-2	The viewpoint applies only to a specific business domain

For example, Lankhorst et al. [13] proposed an infrastructure viewpoint to describe the hardware and *software* infrastructure upon which the application layer depends. In EA debt analysis, this viewpoint can provide insight into divergent solutions by representing the relevant software and hardware infrastructure on which the application layer relies. Furthermore, E. Arkin and B. Tekinerdogan [27] proposed an algorithm decomposition viewpoint to describe a decomposition of the algorithm into different sections, which can be either serial or parallel and are used to analyze algorithms. This viewpoint is not eligible for EA debt analysis because it describes low-level software information that is not relevant to enterprise architecture-level concerns.

Templates for Documenting Viewpoints In the primary studies included in our systematic mapping studies, many of the studies documented the viewpoints by using or adapting the viewpoint templates proposed in standards such as IEEE 1471-2000 [28], ISO/IEC 42010:2007 [29], and ISO/IEC/IEEE 42010:2011 [30]. Furthermore, the standard proposed in [30] is the revision of [28, 29]. The elements proposed in each standard to document the viewpoints are shown in Table 6.

3.4 Evaluation

To evaluate the idea of analyzing EA debt by applying viewpoints, a qualitative evaluation was employed. Further, a questionnaire was created, and people associated with EA debt in industry and academia were asked to fill out the questionnaire for evaluation. Furthermore, the evaluation and its results are discussed in Sect. 5.

3.5 Demonstration

As discussed, by applying viewpoints, stakeholders can have insight into EA debt. For example, Alexopoulou et al. [31] proposed a business process viewpoint to describe the high-level structure and composition of one or more business processes. It describes actions included in the business process, roles of actors, events that govern sequence and execution of actions, and rules which specify the different paths of actions such as routes, resources used, and produced during execution of the

Table 6 Templates for documenting viewpoints

Source	Elements in viewpoint template
IEEE 1471-2000 [28] ISO/IEC 42010:2007 [29]	Viewpoint name Stakeholders Concerns Language, modeling techniques, or analytical methods to be used in constructing views Source for the viewpoint Additionally, viewpoints may contain the following information: Formal or informal consistency and completeness tests to be applied to the models making up an associated view, evaluation, or analysis techniques to be applied to the models, and heuristics, patterns, or other guidelines to assist in the synthesis of an associated view
ISO/IEC/IEEE 42010: 2011 [30]	Viewpoint name: It is the name of the viewpoint Viewpoint overview: A brief description of the viewpoint Concerns: The concerns framed by the viewpoint Typical stakeholders: The list of stakeholders for the viewpoint Model-kind: It consists of conventions for the type of modeling, and it can be documented by either or the combination of the following methods: by creating a metamodel, providing a template, or by language definition or reference to the existing language Correspondence rules: These are the rules that enforce relations in or among architecture descriptions Operations on views: These are the methods that can be applied to views, and they can be creation, interpretive, analysis, design, or implementation methods Examples: An example of a viewpoint Notes: It is the additional information about viewpoint Sources: It refers to the source of viewpoint

action [31]. In EA debt analysis, this viewpoint can assist stakeholders in decision-making by providing insight into the high-level structure and composition of business processes that are sub-optimal and consequently contain debt.

3.6 Communication

To communicate the findings of this study to researchers and practitioners associated with the area of EA debt, the viewpoints applicable for EA debt analysis gathered as a result of the systematic mapping study have been documented online in a public repository in [20].

4 Results

As the result of the systematic mapping study, a total of 216 viewpoints are extracted from 59 primary studies that were selected for final review. Furthermore, the viewpoints gathered from literature are analyzed for their applicability for EA debt analysis. Consequently, considering inclusion criteria, 53% of viewpoints were found applicable for EA debt analysis, and 47% of viewpoints, considering exclusion criteria, were found inapplicable for EA debt analysis.

Viewpoints that are found applicable for EA debt analysis are then categorized into purpose-based and content-based categories. Steen et al. [3] defined three categories to categorize EA viewpoints based on the purpose they serve. These categories are details, deciding, and informing viewpoints. As well, Steen et al. [3] defined three more categories to categorize EA viewpoints based on content their view may contain. These categories are details, coherence, and overview viewpoints. Moreover, these categories have been redefined in this study, so that viewpoints for EA debt can be categorized into these categories. The redefined purpose-based categories for EA debt viewpoints are as follows:

Designing viewpoints provide detailed insight into the solution designs that are considered as deviations from the enterprise architecture standards and principles. Typical stakeholders for such viewpoints are architects, business process designers, software developers, or others who are in charge to evolve/implement the solution designs discussed. Such viewpoints consist of solution design representations such as UML diagrams, BPMN models, and flowcharts.

Deciding viewpoints provide insight into relations between the divergent solutions identified across domain architectures and the factors considered in decision-making processes, usually by employing analytical techniques, projections, and model intersections. Such viewpoints consist of representations in the form of reports, lists, cross-reference tables, and landscape maps. Typical stakeholders of such viewpoints are managers, CIOs, CEOs, or others who have influence on the planning of future IT transformation road-map.

Informing viewpoints provide insight into the development of divergent solutions within a particular area of interest and relevant interpretations of it, which may trigger discussions and further actions. Such viewpoints consist of overviews such as dashboards, history overviews, interpretations, and recommendations. Typical stakeholders of such viewpoints are application owners, business users, business process owners, or others who use or manage particular domain architectures.

Furthermore, the redefined content-based categories for EA debt viewpoints are as follows:

Details viewpoints focus on describing divergent solutions from a single aspect or area of the enterprise architecture. For example, such a viewpoint may represent divergent solutions from a particular business capability or focus only on the financial aspect. Such viewpoints may contain high-fidelity representations such as UML diagrams and business process diagrams. Typical stakeholders for such viewpoints are software engineers, business process owners, subject matter experts, or others whose interests are focused on a particular area or aspect.

Coherence viewpoints focus on describing divergent solutions from a set of related aspects or areas of the enterprise architecture. Such viewpoints could be used for dependency or impact analysis of divergent solutions within a portfolio of enterprise architecture entities. Typical stakeholders for these viewpoints are operational managers, application managers, or others whose interests span across multiple aspects or areas.

Overview viewpoints provide a holistic view of divergent solutions across the enterprise architecture. These viewpoints consist of overarching representations, such as landscape maps. Typical stakeholders for such viewpoints are enterprise architects, CEOs, CIOs, and others who have a stake in managing IT investment portfolio.

Each viewpoint found applicable for EA debt analysis can provide insight into EA debt by focusing on specific aspects and areas of architecture. For example, Lankhorst et al. [13] proposed an implementation and deployment viewpoint to show the deployment of one or more applications on the technical infrastructure. In EA debt analysis, this viewpoint can provide insight into divergent solutions by showing the deployment of relevant applications on technical infrastructure. As it shows the dependency between applications and technical infrastructure, based on the content it contains, this viewpoint can be categorized as coherence viewpoints. As well, this viewpoint can provide insight into the design of divergent solutions by showing the deployment of applications, and such information can assist stakeholders in decision-making. So, based on the purpose this viewpoint can serve in EA debt analysis, it can be categorized as designing and deciding viewpoints.

Further, the business process viewpoint proposed by Alexopoulou et al. [31] describes the high-level structure and composition of one or more business processes. In the context of EA debt analysis, this viewpoint can provide insight into divergent solutions by focusing on business processes that are sub-optimal. Since this viewpoint focuses solely on business processes, so based on the content it contains, it can be categorized as details viewpoints. Also, this viewpoint can provide insight into the design of divergent solutions by describing the high-level structure and composition of business processes. This information can assist stakeholders in decision-making. Therefore, based on the purpose this viewpoint can serve in EA debt analysis, it can be categorized as details and deciding viewpoints.

Furthermore, P. Clements and L. Bass [17] proposed a business goals viewpoint for capturing architecturally significant business goals and architecture strategies for achieving them. In EA debt analysis, this viewpoint can provide an overview of EA debt to stakeholders by providing insight into business goals that are not met as a result of sub-optimal solutions. Based on the content this viewpoint contains, it can be categorized as overview viewpoints. In addition, by presenting such information, this viewpoint can assist stakeholders in decision-making and can inform stakeholders about unaccomplished business goals. Therefore, based on the purpose this viewpoint can serve in EA debt analysis, it can be categorized as deciding and informing viewpoints.

Moreover, the total number of viewpoints found applicable for EA debt analysis is very large, and due to space limitations, they cannot be presented here. Therefore, by employing the viewpoints documentation template proposed in [30], the applicable viewpoints have been documented online in [20].

5 Evaluation

A solution artifact's utility, quality, and efficacy are rigorously demonstrated through well-executed evaluation techniques [32]. Further, various essential quality aspects of the artifact such as functionality, completeness, accuracy, performance, reliability, usability, and adaptability can be assessed in the evaluation [32].

To evaluate the idea of viewpoints-based analysis of EA debt, the selection of viewpoints from literature for EA debt analysis, and the categorization of the viewpoints applicable for EA debt analysis, a qualitative evaluation was employed. Furthermore, a questionnaire was created, and the people associated with the area of EA debt (researchers, practitioners, and master's thesis students in this area) were asked to fill out the questionnaire for evaluation. Moreover, the questionnaire was created by following the guidelines for designing a questionnaire proposed in [33].

The questionnaire was divided into five sections. Except for the first section, each question in all sections was close-ended. The scale used in the form of options for questions about the familiarity of evaluators with concepts of EA debt and architectural viewpoints in the background of the evaluator section is as follows:

Scale: 1 = Not Familiar, 2 = Slightly Familiar, 3 = Moderately Familiar, 4 = Familiar, 5 = Very Familiar

For the question about the capacity in which evaluator is associated with EA debt, four options were provided which are as follows: As EA debt Researcher, As EA debt Practitioner, As Masters Student (Thesis in EA debt area), and others option.

Furthermore, the scale used in form of the options for the questions in the remaining sections is as follows:

Scale: 1 = Strongly Disagree, 2 = Disagree, 3 = Neither Agree nor Disagree, 4 = Agree, 5 = Strongly Agree

The first section consisted of the background of the evaluation. In this section, the relevant concepts and the purpose for evaluation were discussed briefly. The second section consisted of the questions regarding the background of evaluators. It included questions about their familiarity with the concept of EA debt, familiarity with the concept of architectural viewpoints, and a question regarding the capacity in which they are associated with EA debt. Regarding responses to this section, on average, all the evaluators were familiar with concepts of EA debt and architectural viewpoints. Further, a total of six people participated in the evaluation. Furthermore, 33% of evaluators were EA debt researchers, 16.7% were EA debt practitioners, and 50% of evaluators were master's students with a thesis in the area of EA debt. After this, the third section consisted of questions regarding the evaluation of viewpoints-based analysis of EA debt. The purpose behind the questions in this section was to evaluate if the viewpoints can be used for EA debt analysis and whether the predefined viewpoints from literature can inspire the development of viewpoints for EA debt. Regarding the response of evaluators to this section, on average all evaluators agreed to both the notions. Furthermore, the fourth section consisted of questions about the selection of viewpoints from literature for EA debt analysis. This section consisted of questions to evaluate the selection and rejection of various viewpoints from literature for EA debt analysis. The average response of the evaluators for both selection and rejection of viewpoints from literature was toward agreeing side. The last section consisted of questions regarding the definitions of categories for viewpoints that are found applicable for EA debt analysis. Furthermore, the average response of evaluators to the definition of categories was also toward the agreeing side.

Based on the responses to each section in the questionnaire, it is concluded that the results of the overall evaluation are positive. Moreover, questions for evaluation, along with their respective responses, are presented in the Appendix.

6 Discussion

In this section, first, the answers to the research question and pertinent sub-research questions are discussed. After this, the implications of the results to the researchers and practitioners are discussed. Then, the threats to the validity of the study are discussed.

6.1 Answer to Research Questions

The main research question has been divided into three sub-research questions. Therefore, by answering each sub-research question, the main research question (RQ-1: What are the viewpoints that could be used in the context of EA debt analysis?) is answered.

RQ-1.1: What are the existing viewpoints in the areas closely related to EA debt?

As the result of the systematic mapping study carried out to collect viewpoints from the areas of enterprise architecture, software architecture, and technical debt, a total of 216 viewpoints were gathered. However, out of all the viewpoints, 53% of viewpoints were found applicable for EA debt analysis, and the remaining 47% of viewpoints were found inapplicable for EA debt analysis.

Furthermore, viewpoints have been proposed in the literature for a variety of purposes. For example, the primary study [13] proposed an application structure viewpoint along with other viewpoints to describe EA. This viewpoint describes the structure of one or more applications or components. In the context of EA debt analysis, this viewpoint can provide insight into divergent solutions by focusing on the structure of relevant applications.

RQ-1.2: How to derive viewpoints that are relevant for analyzing EA debt from some existing viewpoints used in related areas of study?

To analyze the applicability and inapplicability of viewpoints gathered as a result of the systematic mapping study, inclusion and exclusion criteria were established which are shown in Table 4 and Table 5, respectively. Furthermore, by employing inclusion criteria, the viewpoints from the literature that can provide insight into the design of EA, relations in cross-domain architectures, and viewpoints that can assist in informing stakeholders about the EA have been included to be used in the context of EA debt analysis. As well, the viewpoints from the literature which focus on a specific aspect or area of architecture, viewpoints that focus on more than one aspect or area of architecture, and viewpoints that provide a holistic view of architecture by focusing on multiple aspects and areas of architecture have also been included to be used in the context of EA debt analysis. Moreover, by applying exclusion criteria, viewpoints from the literature which describe low-level software architecture details are not included to be used for EA debt analysis. This is because such viewpoints frame concerns that are not relevant to enterprise architecture-level concerns. Furthermore, viewpoints that are specific to certain business domains (such as digital signal processing software, real-time and embedded systems, physical protection systems) are also not included to be used for EA debt analysis. This is because such viewpoints are specific to a certain business domain.

RQ-1.3: How to document a viewpoint of EA debt?

In the primary studies that were selected for final review in the systematic mapping study, the majority of the viewpoints are documented adapting or using viewpoint documentation templates proposed in IEEE 1471-2000 [28], its revision ISO/IEC 42010:2007 [29], and ISO/IEC/IEEE 42010:2011 [30] (revision of IEEE 1471-2000 and ISO/IEC 42010:2007).

Furthermore, a variety of viewpoints in literature are documented by employing the viewpoints documentation template proposed in [30]. For example, viewpoints proposed for documenting architectural assumptions in [34], viewpoint proposed for describing the context of software systems in [14], viewpoint proposed to describe variability in enterprise software systems in [35], viewpoints to describe architectural decisions proposed in [36], viewpoint proposed to frame concerns regarding changes in enterprise in [37], and viewpoints for documenting architectural technical debt proposed in [19] are documented by employing viewpoints documentation template proposed in [30].

This template has been used to document a variety of viewpoints in literature, including viewpoints for architectural technical debt. Therefore, this template can also be utilized to document viewpoints for EA debt analysis. Consequently, the viewpoints from the systematic mapping study which are applicable for EA debt analysis have been documented in the online repository using the template defined in [30].

6.2 Implications of Results

The idea of viewpoints-based analysis of EA debt and the collection of viewpoints for EA debt analysis benefits both academia and practitioners. The contribution of this work has an implication for research that it provides an initial catalog for EA debt viewpoints which can be extended further by adding more viewpoints. Furthermore, it has an implication for practice that it allows practitioners to use viewpoints to analyze EA debt.

6.3 Threats to Validity

This study consisted of the following threats to validity:

Internal Validity The internal validity is concerned with the extent to which the study's design and execution are likely to prevent systematic error [26]. It is critical to include relevant papers to ensure the study's completeness [38]. Therefore, to include the majority of the relevant papers from the literature, it was decided to conduct an automatic search for primary studies in multiple essential digital databases. Further, in automatic searches, inappropriate or incomplete search terms may result in the exclusion of relevant studies [38]. To mitigate this threat, trial searches were performed, and the search terms were improved eventually. Furthermore, the validity of the answer to a research question depends on the accuracy of the data items extracted [38]. From each study, information of every entity that has been used to document viewpoints is extracted and stored in an excel spreadsheet.

External Validity The external validity is concerned with the extent to which the findings of the study are applicable outside of the study [26]. The viewpoints collected as a result of the systematic mapping study could be applied for EA debt analysis in different enterprises. However, it may be required to adapt the viewpoints to the organizational context of the enterprise.

7 Conclusion and Future Work

EA debt addresses both the business and technical aspects of debt, giving a holistic overview of deviations in the EA. Furthermore, EA debt when accumulated results in negative effects on the business of an enterprise. EA debt has many different manifestations, and it can come from any portion of the architecture that corresponds to one of the four levels of EA. Consequently, its effects may be evident in other layers of the architecture as well.

An enterprise consists of several stakeholders, each of whom is interested in different aspects and areas of architecture. Therefore, in the context of EADM, we propose the use of viewpoints for the analysis of EA debt. By defining and employing viewpoints, stakeholders can gain insight into the divergent solutions by focusing on key aspects and areas of EA that are of interest to them. This would allow stakeholders to make informed decisions about EA debt. As EA debt is a relatively new area, therefore, there are no viewpoints that are specifically proposed for this area. So, in this paper, a systematic mapping study is carried out to gather viewpoints from areas closer to EA debt. After this, each viewpoint is analyzed for its applicability for EA debt analysis. Further, each viewpoint applicable is categorized into redefined categories, and each viewpoint is documented in an online repository that may serve as a knowledge base for viewpoints for EA debt analysis. Furthermore, the idea of viewpoints-based analysis is evaluated by people associated with EA debt in academia and practice. The overall evaluation is positive. Hence this advocates the usage and efficacy of viewpoints-based analysis of EA debt.

Moreover, further research is suggested in the future for establishing a methodology and tool support for viewpoints-based EA debt analysis. As well, the practical applicability of viewpoints-based EA debt analysis requires further evidence of usage, and more research is suggested for the integration of viewpoints-based EA debt analysis in the EADM framework.

Appendix

Q1: How familiar are you with the concept of EA debt? Response: Very Familiar: 33.3%, Familiar: 50%, Moderately Familiar: 16.7%
Q2: How familiar are you with the concept of architecture viewpoints? Response: Very Familiar: 50%, Familiar: 33.3%, Slightly Familiar: 16.7%
Q3: In what role are you familiar with EA debt? Response: As EA Debt Researcher: 33.3%, As EA Debt Practitioner:16.7%, As a Student (Thesis in EA Debt area): 50%
Q4: Do you agree that current architectural and debt-related viewpoints from literature can inspire the development of EA debt viewpoints? Response: Agree: 83.3%, Strongly Agree:16.7%

(continued)

Q5: Do you agree that viewpoints can assist in the analysis of the design of solutions related to EA debt? Response: Strongly Agree: 66.7%, Agree: 16.7%, Strongly Disagree: 16.7%
Q6: Do you agree that viewpoints can assist in making decisions about EA debt by presenting information required for making decisions? Response: Strongly Agree: 50%, Agree: 33.3%, Neither Agree nor Disagree: 16.7%
Q7: Do you agree that by presenting stakeholder-relevant interpretations of EA debt, viewpoints can inform stakeholders about EA debt? Response: Strongly Agree: 66.7%, Neither Agree nor Disagree: 33.3%
Q8: Do you agree that viewpoints can assist in doing a detailed analysis of EA debt? Response: Strongly Agree: 33.3%, Agree: 16.7%, Neither Agree nor Disagree: 50%
Q9: Do you agree that viewpoints can assist in dependency and/or impact analysis of EA debt? Response: Strongly Agree: 16.7%, Agree: 50%, Neither Agree nor Disagree: 16.7%, Disagree: 16.7%
Q10: Do you agree that viewpoints can provide a holistic view of EA debt? Response: Agree: 33.3%, Neither Agree nor Disagree: 33.3%, Disagree: 16.7%, Strongly Disagree: 16.7%
Q11: Viewpoints from the literature that can provide insight into enterprise architecture design have been included to be used in the context of EA debt analysis. Do you agree that in EA debt analysis, such viewpoints, can provide an insight into the design of divergent solutions? Response: Strongly Agree: 33.3%, Agree: 50%, Neither Agree nor Disagree: 16.7%
Q12: Viewpoints from the literature that can provide insight into relations in cross-domain architectures have been included to be used in the context of EA debt analysis. Do you agree that in EA debt analysis, such viewpoints can provide insight into relations among divergent solutions in cross-domain architectures? Response: Agree: 50%, Neither Agree nor Disagree: 33.3%, Disagree: 16.7%
Q13: Viewpoints from the literature that can assist in informing stakeholders about the enterprise architecture have been included to be used in the context of EA debt analysis. Do you agree that in EA debt analysis, such viewpoints can be useful to inform stakeholders about divergent solutions? Response: Strongly Agree: 16.7%, Agree: 66.7%, Neither Agree nor Disagree: 16.7%
Q14: Viewpoints from the literature that focuses on a specific aspect or area of enterprise architecture have been included to be used in the context of EA debt analysis. Do you agree that in EA debt analysis, such viewpoints can provide insight to divergent solutions by focusing on a specific aspect or area of enterprise architecture? Response: Strongly Agree: 33.3%, Agree: 50%, Neither Agree nor Disagree: 16.7%
Q15: Viewpoints from literature which focus on more than one aspect or area of architecture have been included to be used in the context of EA debt analysis. Do you agree that in EA debt analysis, such viewpoints can provide insight into divergent solutions by focusing on more than one aspect or area of enterprise architecture? Response: Strongly Agree: 16.7%, Agree: 50%, Neither Agree nor Disagree: 16.7%, Disagree: 16.7%
Q16: Viewpoints from the literature that provides a holistic view of architecture by focusing on multiple aspects and areas of architecture that have been included to be used in the context of EA debt analysis. Do you agree that in EA debt analysis, such viewpoints can provide a holistic view of divergent solutions across enterprise architecture? Response: Strongly Agree: 16.7%, Agree: 50%, Neither Agree nor Disagree: 16.7%, Disagree: 16.7%
Q17: Viewpoints from the literature that describe low-level software architecture details (such as parallelism, threads, organization of source code into modules, etc.) are not included to be used for EA debt analysis. Do you agree that such viewpoints describe low-level software details that are

(continued)

not relevant to enterprise architecture-level concerns, and so they are not applicable for analysis of EA debt? Response: Strongly Agree: 16.7%, Agree: 33.3%, Neither Agree nor Disagree: 50%
Q18: Viewpoints from the literature that are specific to certain business domains (such as digital signal processing software, real-time and embedded systems, physical protection systems) are not included to be used for EA debt analysis. Do you agree that such viewpoints should not be considered for EA debt analysis because they apply only to specific business domains? Response: Agree: 33.3%, Neither Agree nor Disagree: 50%, Disagree: 16.7%
Q19: Do you agree that the definition of the designing viewpoints category is appropriate? Response: Strongly Agree: 33.3%, Agree: 33.3%, Neither Agree nor Disagree: 33.3%
Q20: Do you agree that the definition of the deciding viewpoints category is appropriate? Response: Strongly Agree: 50%, Agree: 16.7%, Neither Agree nor Disagree: 33.3%
Q21: Do you agree that the definition of the informing viewpoints category is appropriate? Response: Strongly Agree: 50%, Agree: 16.7%, Neither Agree nor Disagree: 33.3%
Q22: Do you agree that the definition of the details viewpoints category is appropriate? Response: Strongly Agree: 50%, Agree: 16.7%, Neither Agree nor Disagree: 33.3%
Q23: Do you agree that the definition of the coherence viewpoints category is appropriate? Response: Strongly Agree: 33.3%, Agree: 33.3%, Neither Agree nor Disagree: 33.3%
Q24: Do you agree that the definition of the overview viewpoints category is appropriate? Response: Strongly Agree: 50%, Agree: 16.7%, Neither Agree nor Disagree: 33.3%

References

1. N. Banaeianjahromi, K. Smolander, Lack of communication and collaboration in enterprise architecture development. Inf. Syst. Front. **21**(4), 877–908 (2019)
2. M. Zhang, H. Chen, A. Luo, A systematic review of business-IT alignment research with enterprise architecture. IEEE Access **6**, 18933–18944 (2018). https://doi.org/10.1109/access.2018.2819185
3. M. Steen, D.H. Akehurst, H. ter Doest, M.M. Lankhorst, Supporting viewpoint-oriented enterprise architecture, in *Proceedings, Eighth IEEE International Enterprise Distributed Object Computing Conference: EDOC 2004: 20–24 September 2004, Monterey*, (2004)
4. W. Cunningham, The WyCash portfolio management system. SIGPLAN OOPS Mess. **4**(2), 29–30 (1993). https://doi.org/10.1145/157710.157715
5. C. Fernandez-Sanchez, J. Garbajosa, C. Vidal, A. Yague, An analysis of techniques and methods for technical debt management: A reflection from the architecture perspective, in *Software Architecture and Metrics (SAM), 2015 IEEE/ACM 2nd International Workshop on*, (2015)
6. E. Tom, A. Aurum, R. Vidgen, An exploration of technical debt. J. Syst. Softw. **86**(6), 1498–1516 (2013). https://doi.org/10.1016/j.jss.2012.12.052
7. S. Hacks, H. Hofert, J. Salentin, Y.C. Yeong, H. Lichter, Towards the definition of enterprise architecture debts, in *2019 IEEE 23rd International Enterprise Distributed Object Computing workshop: EDOCW 2019: Paris, France, 28–31 October 2019: proceedings*, (2019)
8. Y. C. Yeong, S. Hacks, and H. Lichter, Prioritization of EA Debts Facilitating Portfolio Theory, 2019. [Online]. Available: https://www.swc.rwth-aachen.de/docs/2019_quasoq_hacks.pdf
9. N. Zazworka, C. Seaman, F. Shull, Prioritizing design debt investment opportunities, in *Proceeding of the 2nd working on Managing technical deMTD '11*, (New York, 2011)
10. E. Lim, N. Taksande, C. Seaman, A balancing act: What software practitioners have to say about technical debt. IEEE Softw. **29**(6), 22–27 (2012). https://doi.org/10.1109/MS.2012.130

11. Peter Alexander, Simon Hacks, Jürgen Jung, Horst Lichter, Ulrike Steffens, and Ömer Uludağ, A Framework for Managing Enterprise Architecture Debts-Outline and Research Directions, 2020. [Online]. Available: http://ceur-ws.org/vol-2628/paper1.pdf
12. Z. Li, P. Avgeriou, P. Liang, A systematic mapping study on technical debt and its management. J. Syst. Softw. **101**, 193–220 (2015a). https://doi.org/10.1016/j.jss.2014.12.027
13. M.M. Lankhorst, L. van der Torre, H.A. Proper, F. Arbab, S. Hoppenbrouwers, M. Steen, Viewpoints and visualisation, in *Enterprise Engineering Series*, (2017), pp. 171–214. https://doi.org/10.1007/978-3-662-53933-0_8
14. A. Bedjeti, P. Lago, G.A. Lewis, R.D. de Boer, R. Hilliard, Modeling context with an architecture viewpoint, in *Proceedings – 2017 IEEE International Conference on Software Architecture, ICSA 2017*, (2017). https://doi.org/10.1109/ICSA.2017.26
15. P. Närman, M. Buschle, M. Ekstedt, An enterprise architecture framework for multi-attribute information systems analysis. Softw. Syst. Model. **13**(3), 1085–1116 (2014). https://doi.org/10.1007/s10270-012-0288-2
16. J. Miguens, M. Mira da Silva, S. Guerreiro, A viewpoint for representing costs in enterprise architectures, in *2018 IEEE 20th Conference on Business Informatics (CBI)*, (2018), pp. 10–19
17. P. Clements, L. Bass, The business goals viewpoint. IEEE Softw. **27**(6), 38–45 (2010). https://doi.org/10.1109/MS.2010.116
18. N. Da Silva, M. Da Silva, P. de Sousa, R. Villemaire, R. Villemaire, R. Lagerstrom, S. Halle, A viewpoint for analyzing enterprise architecture evolution, in *Proceedings – 2017 IEEE 21st International Enterprise Distributed Object Computing Conference, EDOC 2017, 2017-January*, (2017). https://doi.org/10.1109/EDOC.2017.13
19. Z. Li, P. Liang, P. Avgeriou, Architecture viewpoints for documenting architectural technical debt, in *Software Quality Assurance: In Large Scale and Complex Software-intensive Systems*, (2016). https://doi.org/10.1016/B978-0-12-802301-3.00005-3
20. Software Construction Lab RWTH Aachen, Enterprise Architecture Debt Viewpoints. [Online]. Available: swc-public.pages.rwth-aachen.de/eadm/ea-debt-viewpoints
21. R. Reboucas de Almeida, Business-driven technical debt prioritization, in *2019 IEEE International Conference on Software Maintenance and Evolution (ICSME), Cleveland, OH, USA, Sep. 2019 – Oct. 2019*, pp. 605–609
22. A. Martini, J. Bosch, An empirically developed method to aid decisions on architectural technical debt refactoring, in *Proceedings of the 38th International Conference on Software Engineering Companion, Austin Texas, 05142016*, pp. 31–40
23. N. Rios, R. Oliveira Spinola, M. Mendonca, C. Seaman, Supporting analysis of technical debt causes and effects with cross-company probabilistic cause-effect diagrams, in *2019 IEEE/ACM International Conference on Technical Debt (TechDebt)*, (2019)
24. K. Peffers, T. Tuunanen, M.A. Rothenberger, S. Chatterjee, A design science research methodology for information systems research. J. Manag. Inf. Syst. **24**(3), 45–77 (2007). https://doi.org/10.2753/MIS0742-1222240302
25. K. Petersen, R. Feldt, S. Mujtaba, and M. Mattsson, Systematic Mapping Studies in Software Engineering (2008)
26. Guidelines for performing systematic literature reviews in software engineering, 2007. [Online]. Available: https://citeseerx.ist.psu.edu/viewdoc/download?doi=10.1.1.117.471&rep=rep1&type=pdf
27. E. Arkin, B. Tekinerdogan, Architectural view driven model transformations for supporting the lifecycle of parallel applications, in *2015 3rd International Conference on Model-Driven Engineering and Software Development (MODELSWARD)*, (2015), pp. 40–49
28. IEEE Recommended Practice for Architectural Description for Software-Intensive Systems: IEEE Std 1471–2000, in IEEE Std 1471-2000, p. 1–30, (2000). https://doi.org/10.1109/IEEESTD.2000.91944
29. ISO/IEC Standard for Systems and Software Engineering – Recommended Practice for Architectural Description of Software-Intensive Systems: ISO/IEC 42010 IEEE Std 1471-2000 First edition 2007-07-15, in ISO/IEC 42010 IEEE Std 1471-2000 First edition 2007-07-15, p. 1–24 (2007). https://doi.org/10.1109/IEEESTD.2007.386501

30. ISO/IEC/IEEE Systems and software engineering -- Architecture description: ISO/IEC/IEEE 42010:2011(E) (Revision of ISO/IEC 42010:2007 and IEEE Std 1471-2000), in ISO/IEC/IEEE 42010:2011(E) (Revision of ISO/IEC 42010:2007 and IEEE Std 1471-2000), p. 1–46, (2011). https://doi.org/10.1109/IEEESTD.2011.6129467
31. N. Alexopoulou, P. Kanellis, M. Nikolaidou, D. Martakos, A holistic approach for enterprise agility, in *Handbook of Research on Enterprise Systems*, (2009). https://doi.org/10.4018/978-1-59904-859-8.ch001
32. A.R. Hevner, S.T. March, J. Park, S. Ram, Design science in information systems research. MIS Quarterly **28**(1), 75 (2004). https://doi.org/10.2307/25148625
33. Guidelines for conducting surveys in software engineering v. 1.1, 2015. [Online]. Available: https://portal.research.lu.se/files/6062997/5463412.pdf
34. C. Yang et al., An industrial case study on an architectural assumption documentation framework. J. Syst. Softw. **134**, 190–210 (2017). https://doi.org/10.1016/j.jss.2017.09.007
35. M. Galster, P. Avgeriou, A variability viewpoint for enterprise software systems, in *2012 Joint Working IEEE/IFIP Conference on Software Architecture and European Conference on Software Architecture*, (2012), pp. 267–271
36. U. van Heesch, P. Avgeriou, R. Hilliard, A documentation framework for architecture decisions. J. Syst. Softw. **85**(4), 795–820 (2012). https://doi.org/10.1016/j.jss.2011.10.017
37. M. Razavian, P. Lago, A viewpoint for dealing with change in migration to services, in *2012 Joint Working IEEE/IFIP Conference on Software Architecture and European Conference on Software Architecture*, (2012), pp. 201–205
38. X. Zhou, Y. Jin, H. Zhang, S. Li, X. Huang, A map of threats to validity of systematic literature reviews in software engineering, in *2016 23rd Asia-Pacific Software Engineering Conference (APSEC)*, (2016)

Detection and Prevention of ARP Cache Poisoning in Advanced Persistent Threats Using Multiphase Validation and Firewall

Muaadh Nasr Al-Mwald, Norziana Jamil, Zul Azri Ibrahim, Zaihisma Che Cob, and Fiza Abdul Rahim

Abstract Protocols define a set of rules that govern the communication between hosts connected via a network. Under normal circumstances, the operation proceeds without incident. However, attackers are always on the lookout for ways to exploit loopholes in protocols. This study aimed to investigate Address Resolution Protocol (ARP) issues and develop a technique to detect and prevent malicious ARP activity and anomalies caused by its various implementations. We propose sending three Internet Control Message Protocol (ICMP) probe packets to each host to validate the new binding, one to the previous binding and the other two to the contemporary binding. ARP packets are used together with these ICMP packets to provide multiphase validation for new entries that have no previous ARP cache entries. The asynchronous nature of the proposed scheme requires no changes to the existing protocol. In addition, the proposed technique uses a host-based firewall to block malicious hosts without affecting the ARP's performance.

Keywords ARP · ARP spoofing attack · ARP cache poisoning · MITM · ICMP Protocol

1 First Section

1.1 Introduction

The majority of cybersecurity incidents are caused by insider attacks [1]. Therefore, weak security implementations and a lack of local network monitoring drive more

M. N. Al-Mwald · N. Jamil (✉) · Z. A. Ibrahim · Z. C. Cob
College of Computing and Informatics, Universiti Tenaga Nasional, Kajang, Selangor, Malaysia
e-mail: Norziana@uniten.edu.my

F. Abdul Rahim
Razak Faculty of Technology and Informatics, Universiti Teknologi Malaysia, Kuala Lumpur, Malaysia

C.-C. Wang, A. Nallanathan (eds.), *Proceedings of the 5th International Conference on Signal Processing and Information Communications*, Signals and Communication Technology, https://doi.org/10.1007/978-3-031-13181-3_12

significant attacks. The architecture of local network connection standards has two types of connection modules. The Open Systems Interconnection (OSI) model is a conceptual framework used to define how layers communicate with each other. The OSI model is categorized into seven distinct layers: the physical layer, data link layer, network layer, transport layer, session layer, presentation layer, and application layer. Each layer provides functionality to the one above it.

The other type of connection module is the Transmission Control Protocol and Internet Protocol Suite, which was developed after the OSI model and consists of four layers: the application layer, transport layer, internet layer, and link layer [2, 3]. In most cases, a network administrator does not concentrate on security devices from the data link layer (Layer 2 of the OSI) because external hackers cannot easily connect to an organization's local area network (LAN) [4]. However, the data link layer of these modules and their addressing scheme are easily exploitable, especially with physical access to the local network. Moreover, when one of the layers fails, no amount of patching at higher layers suffices to eliminate the threat effectively since this form of construction infringes on traditional wisdom to avoid building on shaky foundations. Thus, Layer 2 protection is critical in the network since data is transferred through it.

The OSI model's layers operate independently of one another. Therefore, exemplary security implementations of the network should cover all layers individually. The intrusion detection system, encryption techniques, packet filtering, access lists, and firewalls are used to carry out the security for Layer 3 and the upper layers [5]. Yet the data link layer lacks fine-grain controls to prevent attacks at the data link layer [2]. Furthermore, Layer 2 attacks can result in damage to the whole network. Therefore, enforcing high-level security mechanisms in the upper layers is pointless if the data are compromised at Layer 2. For instance, denial-of-service (DoS) attacks may occur due to ARP cache poisoning/spoofing, resulting in the devices running out of memory and becoming inoperable.

Additionally, man-in-the-middle (MitM) attacks can occur, allowing the attacker to insert himself in the middle of the communication. If an attacker successfully executes a MitM attack, all sensitive information on the network can be stolen. Utilizing various techniques, an intruder can conduct Layer 2 attacks such as MitM and DoS attacks over the web using multiple attacking tools. These attacks can compromise data integrity, confidentiality, and availability [6]. ARP cache poisoning can be in stages two or three of the chain model for advanced persistent threats.

The ARP is a protocol for dynamically mapping IP addresses to hardware addresses, commonly referred to as media access control (MAC) addresses [7]. When a host wants to communicate with another host whose MAC it is unaware of, it broadcasts an ARP request for the MAC associated with the destination's protocol address. Only the host with the corresponding IP address responds unicast with its < IP address, MAC > pair to the sender. Thus, the ARP protocol is critical for LAN communication. However, its inherent flaws are frequently used with other more serious attacks such as MitM and DoS attacks. Attackers can sniff the traffic between two victim hosts using a MitM attack. The attacker uses a DoS attack to prevent a victim host from communicating with others. Therefore, as a consequence, ARP spoofing has become the most dangerous attack type in LANs.

In what follows, we propose a comprehensive method for detecting ARP poisoning. The proposed work focuses on detection and prevention through the use of ARP and ICMP probe packets [8–10], according to the ARP packet that was received. In addition, a local table is kept in the host device to store each host's MAC-IP binding for a more extended period, and it is validated each time a new binding is received. For the contemporary binding received, if the attacker uses an IP of the host that is not in the local cache, the entry is accepted. Additional validation is adopted by sending an ICMP echo to validate and check the host aliveness. The detail of the technique is given in the subsequent sections.

The remainder of the paper is organized as follows: Section 2 describes related works pertaining to detection techniques of ARP poisoning attacks. Section 3 introduces our approach to detect and prevent ARP poisoning attacks. Section 4 presents a simulation that illustrates the proposed approach. Finally, we conclude the paper in Sect. 5.

2 Second Section

2.1 Related Works

Numerous strategies to mitigate ARP spoofing have been previously proposed. However, each has its own set of critical drawbacks [11]. Therefore, based on our research, we classify previously proposed solutions into nine different categories.

2.1.1 Modifying ARP Using Cryptographic Techniques

Ticket-based T-ARP [12] and secure S-ARP [13] use cryptographic solutions that improve the value of executing ARP security by two orders of magnitude over existing protocols [14]. However, the solutions fall short in terms of maintaining each host's public key, upgrading the network stack to configure all hosts, and processing overhead associated with sign generation, verification, and key management. Overall, solutions add cryptographic features to the ARP protocol, but they are incompatible with the standard ARP and have a negative impact on the protocol's performance.

2.1.2 Manually Configuring Static ARP Entries

Static MAC entries using static MAC addresses for all other hosts prevent spoofing, but such an approach is not scalable, and managing all these entries is a full-time job. Moreover, this method can fail miserably if mobile hosts like laptops are introduced into the network irregularly. If an operating system receives a gratuitous ARP packet, it may overwrite a static ARP entry gratuitous ARP (GARP) [15].

2.1.3 Securing Switch Ports

Switch port security or dynamic ARP inspection can be used to prevent ARP spoofing attacks. However, despite its ability to easily prevent ARP spoofing, the cost of implementing such a solution may be prohibitive for most businesses because all old switches should be replaced by the network provider [16].

2.1.4 Kernel-Based Patching

Antidote and Anticap are kernel-based patches that try to protect individual hosts from ARP spoofing. Antidote [17] checks whether the previously learned MAC address is still alive by sending unicasting ARP packets in the network when it receives an ARP reply with a different MAC address. Suppressing the update means adding the offending MAC address to a list of banned addresses. Anticap [18] is a Linux kernel patch that only works in a static environment and is only supported by a few operating systems. Anticap prevents ARP replies with a different MAC address from updating the host's ARP cache. It then drops legal but unnecessary ARP replies, which is against the ARP protocol specification. Both of the above methods require a valid ARP entry in the cache. As a result, the attacker and the victim find themselves in a race. The actual MAC address is banned if the attacker receives a spoofed ARP entry into the node's cache before the existing node.

2.1.5 ARP Spoof Detection and Protection Software

Majumdar A and Raj S [19] produced an ARP attack detection and prevention algorithm using Scapy packet filtering. The algorithm works by comparing the actual MAC address and the ARP packet-sniffed response MAC address. The detection algorithm passively sniffs incoming packets and processes them to see if attacks have occurred. If an attack is detected, a static ARP entry is added to each system on the network. Having static IP and MAC addresses helps prevent spoofing attacks as they can ignore ARP responses.

In contrast, this solution appears to be a satisfactory host-based solution. However, it does not prevent the attacker from continuing to impersonate the victim's MAC address. This is because it's adding a static entry to each entry, and the process will need to be repeated each time a new device joins the network.

Xing et al. [13] used WinPcaP to capture and filter ARP packets. If they differ, an ARP response packet receives an update request and compares the local ARP cache's contents to the proper IP-MAC address pairs. Ramachandran and Nandi [8] investigated ARP and TCP SYN address inconsistencies. They used ARP messages to publicize IP-MAC addresses to create TCP SYN packets. In the absence of ARP attacks, the TCP SYN packet's destination MAC and IP addresses must match the ARP source MAC and IP addresses. However, generating ARP DoS attacks to probe the network generates a lot of traffic on the LAN. Also, this method can only detect ARP attacks, not prevent them.

2.2 *Profile Network Behavior with DHCP Server*

A filter on ARP packet requests on a DHCP-enabled network is used to detect ARP spoofing activity in the proposed model. The source and destination computers' communication details can be gleaned using the filter. A table of ARP communication information is associated with each broadcast. Each computer's MAC address is translated to an IP address in this table [20]. In this way, the proposed model can detect high-performance ARP spoofing attacks. In order to process and detect data efficiently, the ARP control flow model relies on numerous resources. Another problem with this study's packet flow inspection is that it requires a time-consuming and complex process to count how many data packets represent service requests [21]. Each of the three proposed solutions is deficient in some way. The dataset used to verify the Hsian solution's findings was not made public, whereas the proposed approach did. Kaur's conclusions indicated that Naive Bayes had the lowest FAR and was 93% accurate, making it the solution's most precise prediction method, which is not the best accuracy for machine learning models. And Ma et al.'s solution identified the attack solely based on four characteristics, without any experimental evidence.

2.2.1 The Use of Machine Learning to Detect ARP Attacks

Using machine learning, Hsiao et al. proposed an ML approach for ARP attack detection. He performed ARP attack detection using data collected from simulating a network of 256 hosts at the National University of Kaohsiung [22]. Kaur [23] proposed three different methods including (1) signature-based, (2) manual Wireshark packet analysis, and (3) ML. Ma et al. [24] proposed the Bayes way to determine the likelihood of an attack. It used different types of machine learning algorithms to detect attacks based on their probability.

2.2.2 Centralized Server and Client/Server-Based Intrusion Detection System (CSIDS)

With minimal burden, a centralized server can detect and prevent ARP spoofing attacks completely. This method involves three significant modules: a server-perspective model, a host-perspective model, and server-to-server authentication. First, each communicating host must send an IP-MAC pair to the centralized server during the DHCP session in the host module. The MAC-IP match is checked regularly by broadcasting a packet with its own MAC and IP in the field of the source and its IP and ff:ff:ff:ff:ff:ff in the destination field's MAC if an attacker is found by receiving a reply for the above message [22]. Then, the destination host sends the attacker information UDP/TCP message to the centralized server. Other methods proposed for resolving the ARP spoofing problem include the Snort intrusion detection system [23] and the client/server-based intrusion detection system [24]. Those techniques appear to be the best solutions. Still, all require additional hardware, such as a server and other implementation processes, and dedicated personnel to set up the server, which is quite complicated.

2.2.3 Recent Works

Atmojo Y and Susila I [25] proposed using K-NN classification machine learning to develop a model for detecting ARP poisoning attack activity. The solution aimed to deduce the pattern of ARP poisoning attacks by analyzing network traffic. The proposed model extracts activity patterns from ARP poisoning attacks generated during the attack simulation process and labels activity types by trained experts. With a K value of 2 and an accuracy of 98.7%, the proposed model has the highest detection accuracy of the K-NN method, in comparison to a variety of other techniques.

Ahuja N and Singal G [26] proposed classifying benign network traffic from ARP poison and ARP flooding attacks using machine learning (ML). The solution builds a Python application for the SDN controller using Mininet that collects and logs the attack-detection features into a traffic dataset file. This dataset is used to train the ML model. The hybrid CNN-LSTM model outperforms other ML models with a 99.73% accuracy score. During the attack, high CPU and memory usage serve as experimental evidence. The attack detection time of 63,000 microseconds also shows efficiency.

3 Third Section

3.1 Proposed Scheme

We enhance the solution based on the three solutions proposed by Tripathi and Mehtre [27], Pandey [28], and Singh et al. [9]. Our scheme is called a multiphase validation scheme for detecting and preventing ARP cache poisoning. It employs a distributed concept, in which each host maintains the ARP table in the form of a text file containing the host's IP and MAC bindings. The purpose of using the file is to save the binding. Then, the ARP table is updated once the bindings in the incoming packet have been validated. The detection and prevention schemes in our solution are described below.

3.1.1 Detection Scheme

In this scheme, ARP and ICMP broadcast requests are first sent to each ARP entry on the host, updating any existing entries before monitoring incoming traffic. The scheme uses both passive and active techniques to detect and correct the attack immediately, regardless of whether the attack occurs before or after the tool runs.

When an ARP reply is received, the detection host first looks for the binding in its block list. If it is in the block list, the packet will be ignored. Suppose it is not in the block list. In that case, the following three scenarios are possible: it is a new host (Case 1), it is the same host (Case 2), or it is a malicious host (Case 3). First, the local ARP cache bindings are compared to the received host bindings to distinguish between the three scenarios. Next, we check whether there are bindings on the local ARP cache. Case 1 (new host): If no bindings are found in the local ARP cache, this indicates a new host, and the entry is accepted. Three ICMP packets are then sent to verify the host's aliveness. There are two possible situations at this point:

Situation 1: If no response is received from the IP address, this indicates that the host is either unreachable or unavailable, resulting in the deletion of the entry. This method ensures that attacks caused by the so-called ARP storms are prevented.

Situation 2: When an IP response is received, the entry in the ARP cache is updated, and the entry is saved for the next validation session when a second ARP reply is received. Figure 1 shows Case 1 for the binding validation of a new host.

In Case 2 (same host), bindings are found in the local ARP cache. If bindings are obtained without any IP or MAC mismatches, the entry is updated and saved for the next timeout session. Since the ICMP and ARP packets are used to validate the entry, there are three possible scenarios. Fundamentally, the ARP request/reply is a term

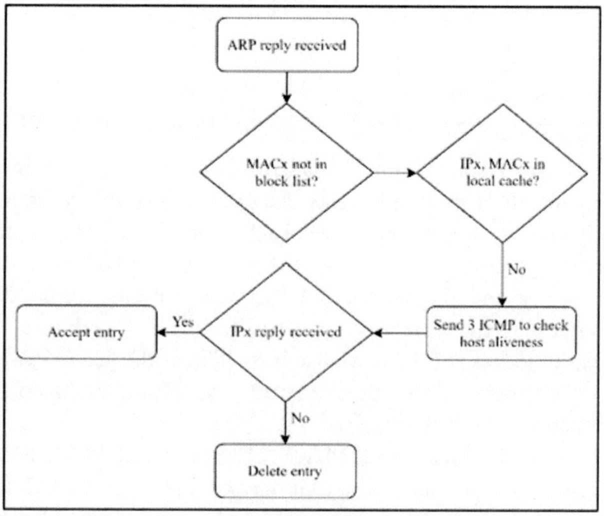

Fig. 1 Case 1 (new host) validation

that shows that the host sends a request/reply and that the host receives a request/reply. If the host wishes to send an ARP request, the host simply broadcasts an ARP request to all devices in the network with its IP, MAC, and the destination IP. If the request is meant for that device, the device will send a unicast message to the device, proving its MAC. Otherwise, it drops the packet. Only devices whose IP matches the destination IP header will respond to the broadcast ARP request. The first scenario is referred to as Case 2.1: When a device receives an ARP request, it checks the header destination IP, i.e., if that request is made for it. If the request is the intended device, it sends a Unicast ARP reply to the sending host with its MAC address. For an ARP reply received, for example, <IPX, MACX>, the device checks if the IPX exists in the local ARP cache. If it is found, then the corresponding MAC entry is checked. If the binding is found to be identical, then it is a normal response, and the local ARP cache table is updated.

The second scenario is referred to as Case 2.2. In this scenario, the bindings received are not identical, indicating a mismatch of the IP or MAC addresses in the received bindings. This indicates the presence of a malicious host. Therefore, entries with mismatches are removed, and three ICMP packets are sent to both IPs in order to obtain the correct bindings and correct the local ARP table. Figure 2 shows the flow of the binding validation mechanism for the same host. The pseudocode below illustrates Case 1 and Case 2.

If ARP packet is received:
If it is an ARP Reply:
Check if Packet MAC is not in the block list:
Check local cache:
If the IP and MAC are not in local cache:
Send 3 ICMP to the packet IP
If reply is received:
Accept entry, else: Delete entry
If the IP and MAC are in the local cache:
Check Mismatch in local cache
If no mismatch exists:
send ARP request to validate the Packet MAC:
If the reply bindings received are the same:
Normal ARP reply & Update entry
If the reply bindings received are not the same:
Malicious host exists, Delete Entries with mismatch, Send ICMP check to packet IP.

Case 3 (malicious host): Bindings are found in the local ARP cache, and a mismatch of the bindings is discovered. There is a mismatch of either the MAC or IP when searching in the local ARP table. Both entries are deleted, and three ICMP probe packets are sent to both the old and new bindings to verify the previous host's viability and the legality of the new binding received. Following that, an ARP request is made in response to the host's IP address in order to obtain the true MAC address of the response host. This is the second validation scheme that verifies and ensures that the received host's MAC address is correct. There are two possible outcomes from this phase.

Case 3.1 (the same MAC): The received MAC address matches the packet's MAC address, indicating that the host bindings are valid, at which point the entry is accepted, or that the received MAC address has already been blocked. Figure 3 shows the malicious host detection.

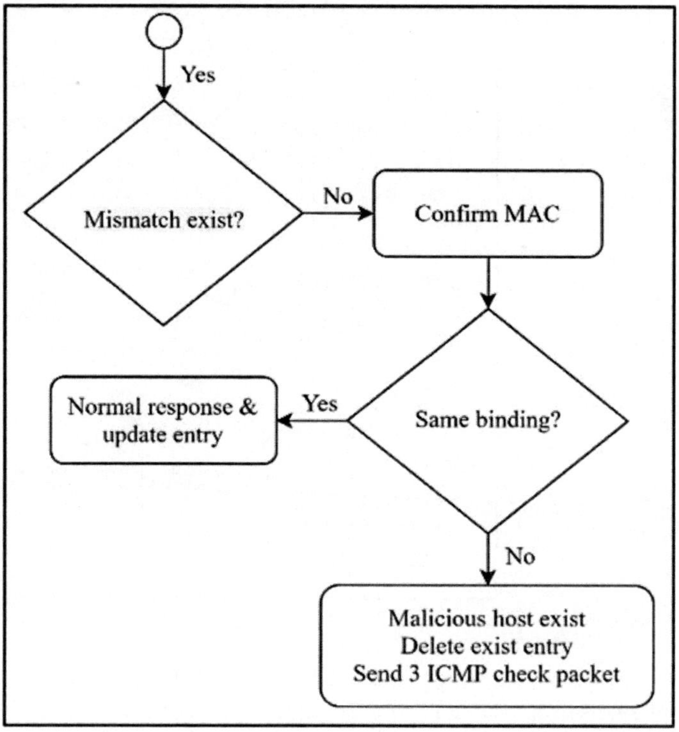

Fig. 2 Case 2 (same host) validation

3.1.2 Prevention Scheme

After an attack is detected, the need for a prevention mechanism is critical. Even though users are notified of the attack's existence, preventing it requires additional work. Even technical users lack the ability to do so if the attack is launched using sophisticated tools such as *bettercap* or Cain and Abel. Each operating system comes pre-installed with a host-based firewall. One of the advantages of our solution is that our detection mechanism makes use of this firewall to prevent attacks once they are successfully identified during the detection process. Following that, the prevention mechanism adds two firewall rules to the host that hosts the tool, preventing incoming and outgoing traffic to the MAC discovered in the packet at OSI Layer 2 (data link). Additionally, two firewall rules are added using IP tables to prevent traffic from reaching the packet MAC at OSI Layer 3 (network). These firewall rules ensure that the attacker does not acquire any traffic and that the tools used to launch an ARP spoofing attack fail.

Case 3.2 (MAC mismatch): When the host responds with a different MAC address and the MAC address is not on the block list, the entry is deemed invalid. Thus, the packet is dropped and deleted from the local ARP cache. At this point, the

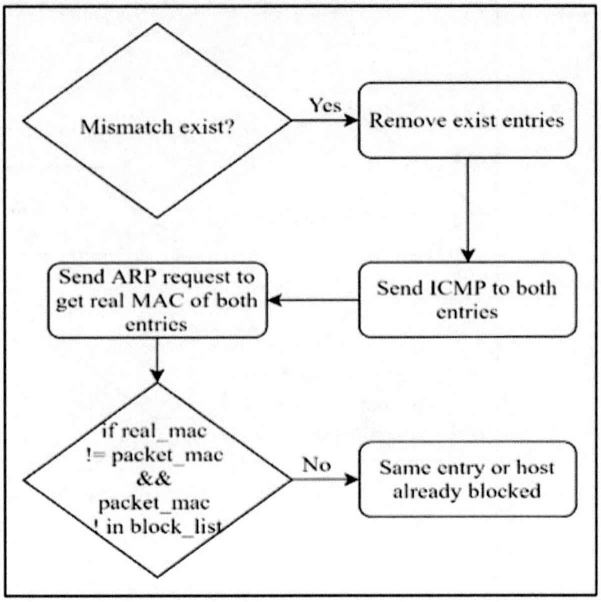

Fig. 3 Case 3.1 (malicious host) detection

detection mechanism has successfully spotted the attack. However, dropping packets is insufficient against some advanced tools. These tools can generate a large amount of ARP spoofs per minute, allowing them to bypass most detection tools. Therefore, the prevention mechanism will immediately add rules to the firewall to ensure that the attacker does not capture any traffic, and the tools used to launch an ARP spoofing attack will fail at this point. Finally, the user is notified of the attack, information regarding the attack is logged into the OS, and an email is sent to the host to report the incident. Figure 4 depicts the prevention mechanism flow.

3.1.3 Attack Report and Alert

In the event of an attack, two types of reports are generated:

1. Logging the attack incident into system logs.
2. Sending an email to the user. Following that, an alarm is triggered.

The pseudocode for Case 3.1 (malicious host) is given below, as well as that for the report and alert mechanism.

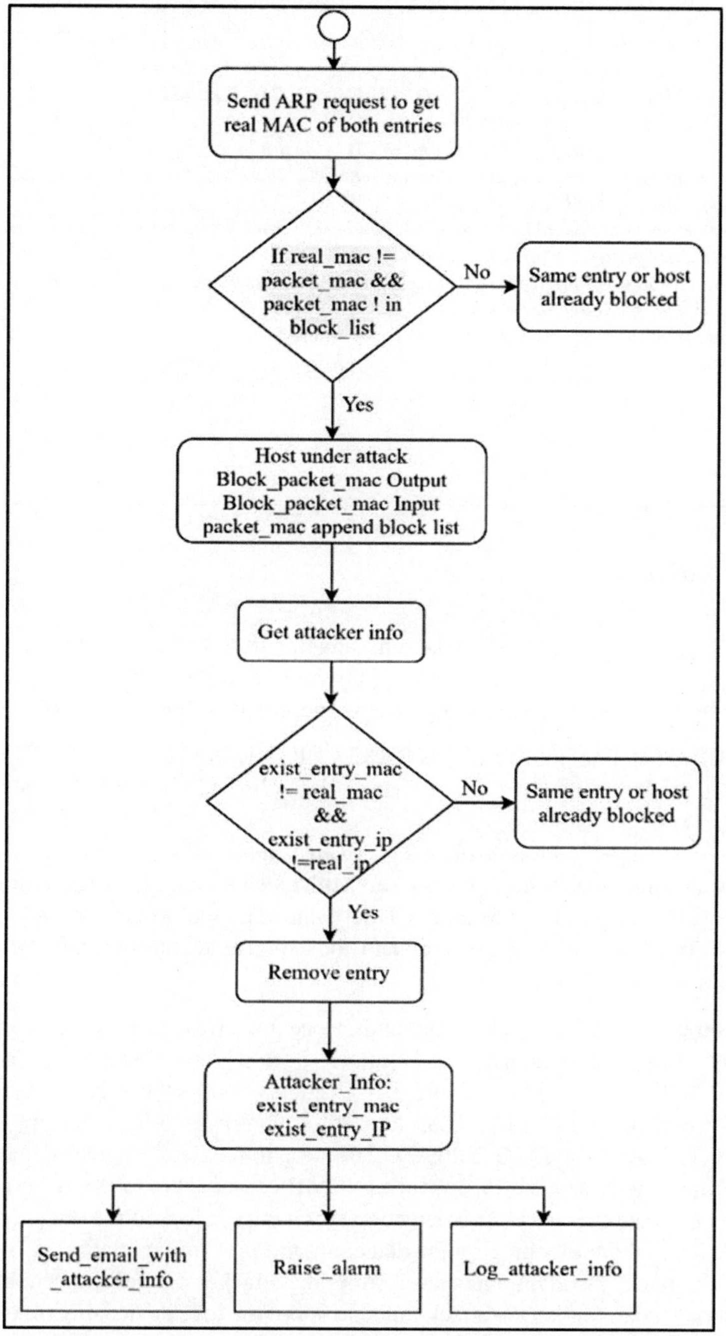

Fig. 4 Case 3.2 (malicious host) attack prevention

If a mismatch exists:
remove existing entries, send three ICMP packets to both entries, send an ARP request to both entries to get the real MAC of the entries.
If the real MAC is an equal packet MAC or the packet MAC in the block list:
The same entry or host is already blocked.
If the real MAC is not equal packet MAC and the MAC is not in the block list:
the host is under attack, block output to the packet MAC, block input from the packet MAC, and append packet MAC to the block list.
If exist_entry_MAC is not equal to the real MAC and exist_entry_IP that is not equal to the real IP:
remove the existing entry,
log the existing entry IP and MAC to the log file to report the attacker info, raise an alarm, and send an email with the logged attacker info.
Host already blocked:
Ignore packet.

4 Fourth Section

4.1 Simulation

To conduct an analysis of ARP poisoning attacks, it was necessary to conduct the attacks, experiment with the tools, and capture the generated traffic. All simulations were conducted in a small laboratory environment that comprised the following:

- TOSHIBA Satellite S55t-A 16 GB RAM, 2.40 GHz Intel Core (TM) i7-4700MQ (Ubuntu x86 64) in the role of a user defending against ARP attacks with *arpShield*
- Attacker: Raspberry Pi 4 (8GB RAM) (Kali Linux), which plays the role of an ARP spoofing attacker that launches an MitM attack using the *bettercap* tool
- Router TP-Link AC1350 Wireless Dual Band Gigabit Router (router, switch, access point), connecting the user and the attacker to the Internet via a local wireless network

The simulation of the above-defined procedure was carried out using the *bettercap* tool to launch the attack from a Raspberry PI (IP: 192.168.0.107, MAC: dc:a6:32:e8:12:c4) against both the user running the *arpShield* tool (IP: 192.168.0.106, MAC: a4:db:30:7a:1d:37) and router (IP: 192.168.0.1, MAC: 84:d8:1b:ff:d0:e7). The tool immediately detected the attack and identified the attacker info. It sent an ICMP echo packet to the IP in order to correct and update the entry. The detection mechanism blocked the attacker MAC. Figure 5 shows the tool's process for detecting and preventing attacks.

Once an attack was identified successfully, the attacker's bindings were blocked, and the tool used to launch the attack failed to spoof the user device. Figure 6 depicts the result from *bettercap*.

Figure 7 shows the firewall rules that were added to the host device after the attack was detected.

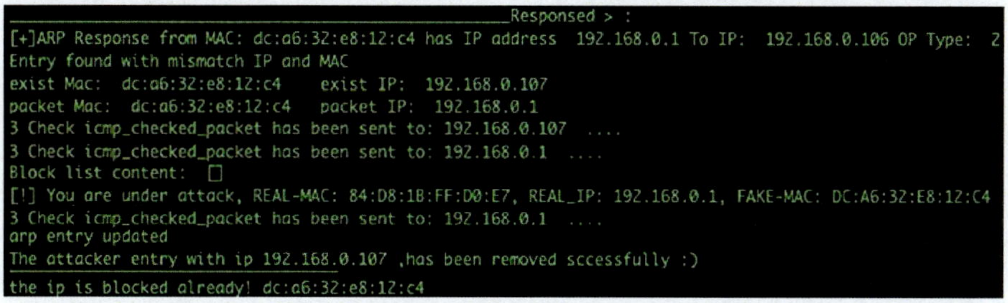

Fig. 5 Attack detection from the tool in the user device

Fig. 6 Bettercap's failure against the arpsheild tool

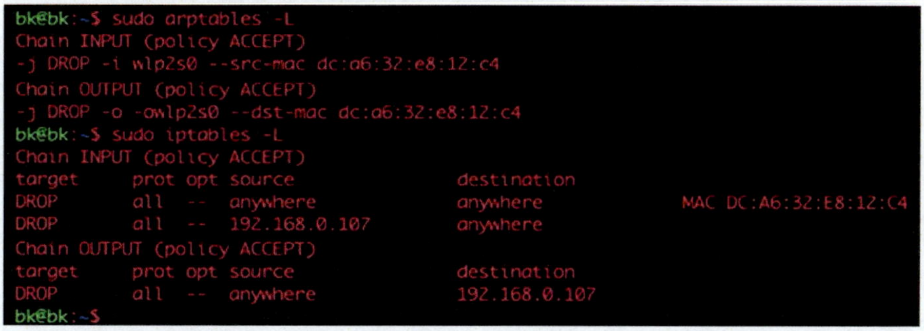

Fig. 7 The added rules to arptables and iptables after the attack is detected

We used *bettercap*, Cain and Abel, and a Python script on three devices to simulate an ARP attack and capture the traffic generated for 5 min. Figure 8 shows the tool's low impact on the CPU when monitoring and defending against attacks: It used less than 35% of the CPU utilities. The tool also consumed no CPU units when no ARP requests were received, allowing it to be run at system start-up for long-term use with no impact on CPU performance.

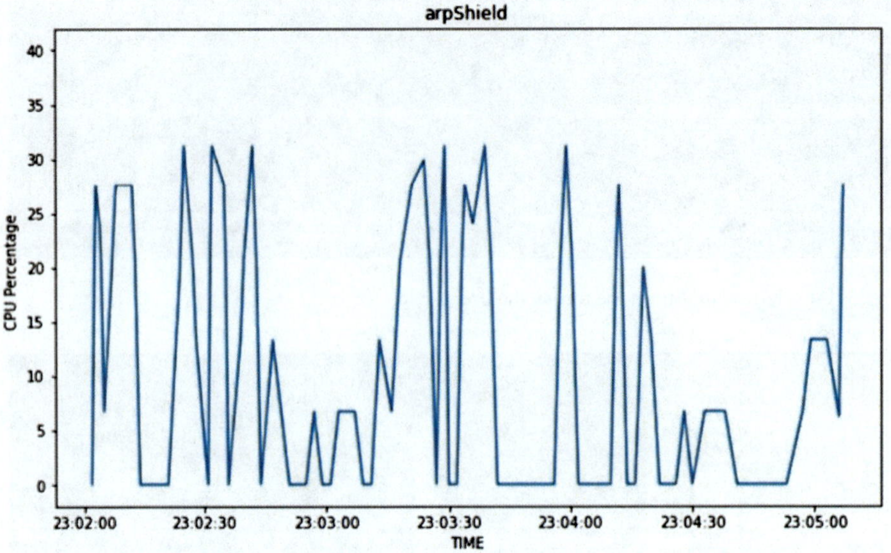

Fig. 8 CPU performance

5 Fifth Section

5.1 Conclusion

This paper proposed a new technique to detect ARP poisoning attacks with the capability of preventing them automatically through the three main steps detailed above. Our technique can also defend against other types of attacks, including MitM, session hijacking, and impersonation of the host. The binding can be kept using the local ARP table, and the attacker's attempt can be easily identified by sending ICMP probe packets to both the previous and new hosts. If the host is still alive in the network, an ARP request will be sent to determine all the hosts claiming to hold that IP address. An ARP request is sent even for new bindings. There is another phase validation using ARP with ICMP packets, even if it is alive and the attacker sends a spoofed reply. Then, the prevention mechanism blocks the attacker's bindings, permanently preventing the attacker from launching the attack. This is an asynchronous scheme that eliminates the need for periodic monitoring. We implemented the proposed technique in a small laboratory to provide a proof of concept for the result. Our experiment showed that the technique detects ARP poisoning attacks quickly and is able to prevent future attacks. Future work will involve expanding our work by improving a few sites and enhancing the technique used to support the implementation of mobility technology.

Acknowledgments This research is supported by the Ministry of Higher Education Malaysia and Universiti Tenaga Nasional (UNITEN), under the Transdisciplinary Research Grant Scheme (TRGS) of Grant No. TRGS/1/2020/UNITEN/01/1/2.

References

1. Schwab W, Poujol M. The State of Industrial Cybersecurity 2018. 2018
2. Networks-Baseline. OSI model vs TCP/IP model – Route XP Private Network Services (no date)
3. N. Malviya, *The OSI Model and TCP/IP Model – Infosec Resources* (INFOSEC, 2020)
4. E. Vyncke, C. Paggen, *LAN Switch Security: What Hackers Know About Your Switches* (Cisco Press, 2007)
5. T. OConnor, *Detecting and Responding to Data Link Layer Attacks* (SANS Institute, 2010)
6. H. Altunbasak, S. Krasser, H. Owen, J. Sokol, J. Grimminger, H.P. Huth, Addressing the weak link between layer 2 and layer 3 in the internet architecture, in *29th Annual IEEE International Conference on Local Computer Networks*, (2004)
7. D.C. Plummer, *RFC 826 – An Ethernet Address Resolution Protocol: Or Converting Network Protocol Addresses to 48.bit Ethernet Address for Transmission on Ethernet Hardware* (Network Working Group, 1982)
8. G. Jinhua, X. Kejian, ARP spoofing detection algorithm using ICMP protocol, in *2013 International Conference on Computer Communication and Informatics, ICCCI 2013*, (2013)
9. S. Singh, D. Singh, A.M. Tripathi, Two-phase validation scheme for detection and prevention of ARP cache poisoning, in *Progress in Advanced Computing and Intelligent Engineering*, (Springer, 2019), pp. 303–315
10. P. Arote, K.V. Arya, Detection and prevention against ARP poisoning attack using modified ICMP and voting, in *Proceedings – 1st International Conference on Computational Intelligence and Networks, CINE 2015*, (2015), pp. 136–141
11. C.L. Abad, R.I. Bonilla, An analysis on the schemes for detecting and preventing ARP cache poisoning attacks, in *Proceedings – International Conference on Distributed Computing Systems*, (2007)
12. V. Goyal, R. Tripathy, An efficient solution to the ARP cache poisoning problem, in *Australasian Conference on Information Security and Privacy*, (2005)
13. D. Bruschi, A. Ornaghi, E. Rosti, S-ARP: A secure address resolution protocol, in *Proceedings – Annual Computer Security Applications Conference, ACSAC*, (2003)
14. W. Lootah, W. Enck, P. McDaniel, TARP: Ticket-based address resolution protocol, in *Proceedings – Annual Computer Security Applications Conference, ACSAC*, (2005)
15. Koo J, Ahn S, Lim Y, Mun Y. Evaluation of network blocking algorithm based on ARP spoofing and its application. In: Computational Science and Its Applications – ICCSA 2005. Springer; 2005
16. M.A. Carnut, J.J.C. Gondim, ARP spoofing detection on switched Ethernet networks: A feasibility study, in *5th Simpósio Segurança Em Informática*, (2003)
17. Teterin. Antidote. Security Focus (2002)
18. M. Barnaba. Anticap (2003)
19. A. Majumdar, S. Raj, T. Subbulakshmi, ARP poisoning detection and prevention using scapy. J. Phys. Conf. Ser. **2021** (1911)
20. P. Akhil, B.A. Jose, A Profiling Based Approach to Detect ARP Poisoning Attacks, in *2021 International Conference on Green Energy, Computing and Sustainable Technology, GECOST 2021*, (2021)
21. H.W. Hsiao, C.S. Lin, S.Y. Chang, Constructing an ARP attack detection system with SNMP traffic data mining. ACM Int. Conf. Proc. Series, 341–345 (2009)

22. D. Srinath, S. Panimalar, A.J. Simla, J. Deepa, Detection and prevention of ARP spoofing using centralized server. Int. J. Comp. Appl. **113**, 26–30 (2015)
23. A. Tasneem, A. Kumar, S. Sharma, Intrusion detection prevention system using SNORT. Int. J. Comp. Appl. **181** (2018)
24. Salim H, Li Z. T 1520–9202. (2021)
25. Y.P. Atmojo, I.M.D. Susila, I.B. Suradarma, L. Yuningsih, E.S. Rini, D.P. Hostiadi, *A New Approach for ARP Poisoning Attack Detection Based on Network Traffic Analysis* (2022), pp. 18–23
26. N. Ahuja, G. Singal, D. Mukhopadhyay, A. Nehra, Ascertain the efficient machine learning approach to detect different ARP attacks. Comput. Electr. Eng. **99** (2022)
27. N. Tripathi, B.M. Mehtre, An ICMP based secondary cache approach for the detection and prevention of ARP poisoning, in *2013 IEEE International Conference on Computational Intelligence and Computing Research, IEEE ICCIC 2013*, (2013), pp. 2–7
28. P. Pandey, Prevention of ARP spoofing: A probe packet based technique, in *Proceedings of the 2013 3rd IEEE International Advance Computing onference, IACC 2013*, (2013)

Permeability and Inertia Coefficient of Wire-Mesh Porous Media Estimated by a Darcy-Forchheimer Equation

Suradech Sinjapo, Panuwat Chanmak, and Bundit Krittacom

Abstract The airflow through the wire-mesh screen inside the circular pipe was experimented with to investigate the permeability (K) and the inertia coefficient (F) based on Darcy–Forchheimer's principle. Five pores per inch (PPI) of the SUS304 stainless wire-mesh (i.e., 6, 8, 10, 12, and 14) were examined with the wire diameter (d_w) that was kept at 0.36 mm. From such PPIs, the porosity (ε) yielded 0.866, 0.822, 0.777, 0.733, and 0.688, respectively. Five different stacked thicknesses (H) were conducted in the range of 1.44–7.20 mm. The linear formulation adapted from the Darcy–Forchheimer equation was employed, with $\Delta P = a + bu$ denoting the relationship between pressure drop (ΔP) and velocity (u). The ΔP was measured using a digital manometer by changing u between 0.262 and 1.833 m/s. The experiment revealed that when ε rose, K grew and F dropped. With the impact of H, the opposite situation was obtained—K dropped and F rose. The correlations of K and F were proposed by $K = (1.3425 + 6.2161\varepsilon - 0.4977H) \times 10 - 7$ and $F = 0.6296 - 0.7524\varepsilon + 0.0653H$. To validate the proposed correlations, two available models, consisting of the equation of Gebart and Koponen, were examined

S. Sinjapo
Development in Technology of Porous Materials Research Laboratory (DiTo-Lab), Rajamangala University of Technology Isan, Nakhoracasima, Thailand

Department of Railway System Engineering, Faculty of Engineering and Architecture, Rajamangala University of Technology Isan, Nakhoracasima, Thailand

P. Chanmak
Development in Technology of Porous Materials Research Laboratory (DiTo-Lab), Rajamangala University of Technology Isan, Nakhoracasima, Thailand

Department of Mechanical Engineering, Faculty of Industrial Education, Rajamangala University of Technology Suvarnabhumi, Nonthaburi, Thailand

B. Krittacom (✉)
Development in Technology of Porous Materials Research Laboratory (DiTo-Lab), Rajamangala University of Technology Isan, Nakhoracasima, Thailand

Department of Mechanical Engineering, Faculty of Engineering and Architecture, Rajamangala University of Technology Isan, Nakhoracasima, Thailand
e-mail: bundit.kr@rmuti.ac.th

C.-C. Wang, A. Nallanathan (eds.), *Proceedings of the 5th International Conference on Signal Processing and Information Communications*, Signals and Communication Technology, https://doi.org/10.1007/978-3-031-13181-3_13

for K, and Du Plessis's equation was adopted for F. In comparison, a favorable agreement was reached. As a result, the proposed models were very credible and straightforward to apply.

Keywords Permeability · Darcy–Forchheimer equation · Wire-mesh porous media

1 Introduction

The fluid flow in porous media has received considerable attention because of its importance in the areas of pneumatic–hydraulic applications [1]. The principle of fluid transport within porous media is used in many works—for example, petroleum technology [2, 3], groundwater hydrology [4, 5], micromachining work [6, 7], and fuel cells [8]. For fluid moving at a sufficiently high velocity through porous media, the pressure drop (ΔP) versus flow velocity is given by the Forchheimer equation [9]:

$$\Delta P = \left(\frac{\mu L_0}{K}\right)u + \left(\frac{\rho F L_0}{\sqrt{K}}\right)u_2 \quad (1)$$

where μ, ρ, L_0, K, and F are fluid viscosity, fluid density, testing length, permeability, and inertia coefficient (or Ergun coefficient) [10], respectively. The Forchheimer equation (Eq. 1) modified from the Darcy law [11] has, generally, two terms expressing the viscous and the inertial effects, respectively.

Permeability (K) and the inertia coefficient (F) estimated from the Forchheimer equation are the key to the transmitted ability of structure. For macroscopic phenomena of fluid flow through the porous matrix, the quantity of K and F have perhaps the most important property to describe and to thoroughly understand the flow phenomena. Traditionally, the void structure in a porous matrix, as defined by the porosity (ε), becomes complex, resulting in a convoluted flow pattern inside and across the void [12].

The Kozeny–Carman equation [13, 14] was the most classical equation explained in the relationship between K and porosity (ε). The measurement of K and F for the foam and fibrous structure were conducted by many works [15–17]. In addition, several experimental investigations and numerous numerical analytical details using the lattice Boltzmann technique [18–20] have been undertaken during the previous few decades. Gebart [21] investigated K with a combination method of theoretical, numerical, and experimental studies for which he proposed the functional form of K to ε (porosity). A good application of his function was obtained for $\varepsilon \geq 0.65$. Du Plessis et al. [22] presented a correlation for predicting the F. This equation gave good application in the high porosity range. Koponen et al. [23] applied the lattice Boltzmann method to establish an empirical relationship of K for ε in the range of $0.4 \leq \varepsilon \leq 0.95$.

Recently, the corresponding author and coworker performed experimentally to measure the K of the stainless wire-mesh screen with the porosity range of 0.794–0.943 [24]. The SUS304 stainless wire-mesh screen was tested at four different pores per inch (PPI) (i.e., 4, 8, 10, and 12). The wire diameter did not use the same size. A correlation of K was only proposed but F was not discussed.

Based on previous work [24], this chapter wants to extend the boundary of the investigation. Two PPIs (6 and 14) were added experimentally, but PPI = 4 did not charge because the enhancement performance was insignificant [25, 26]. The wire diameter (d_w) was used with the size kept at 0.36 mm. Thus, five PPIs of the SUS304 stainless wire-mesh (i.e., 6, 8, 10, 12, and 14) were tested by which the porosity (ε) became 0.866, 0.822, 0.777, 0.733, and 0.688, respectively. The wire-mesh screen was prepared as six stacked wire-mesh bundles for which a distance between two adjacent stacked wire-meshes (L) was 80 mm. The stacked thicknesses (H) of wire-mesh porous media organized with five Hs was: 1.44, 2.88, 4.32, 5.76, and 7.20 mm. Finally, two correlations of K and F on ε and H were reported, and the comparison results of a present correlation with available functions were discussed.

Nomenclature
The following nomenclature is used in this chapter:

a	Slope of the line
b	y-axis ($\Delta P/u$) intercept
d_w	Wire diameter, m
F	Inertia coefficient
H	Porous thickness, m
K	Permeability, m^2
L_0	Testing length, m
PPI	Pores per inch
Q	Volumetric flow rate, m^3/s
r	Material diameter, m
R^2	Determination coefficient
T_{in}	Inlet air temperature, °C
u	Fluid velocity, m/s
ΔP	Pressure drop across testing pipe, N/m^2
ε	porosity
ρ	Fluid density, kg/m^3
μ	Fluid viscosity, Pa-s

2 Experimental Setup

2.1 Experimental Diagram

Figure 13.1 shows a schematic diagram of the experiment to evaluate the permeability (K) and the inertia coefficient (F) of wire-mesh porous media attached inside the circular pipe based on the same experimental system of Huchaiyaphum et al.

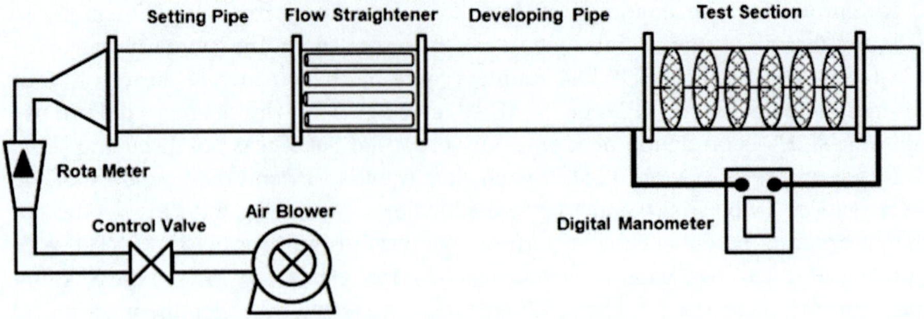

Fig. 13.1 Schematic diagram of the experimental setup [24]

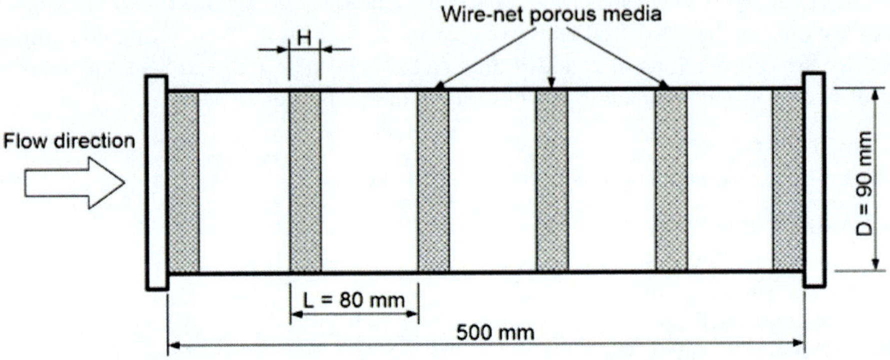

Fig. 13.2 A testing pipe with the installation of wire-mesh porous media [24]

[24]. This test rig is a good general way for monitoring the pressure drop (P) through the pipe because it is a straightforward procedure for determining K and F [27, 28].

To understand in the test section, a testing pipe made of SUS304 stainless steel with an inner diameter of 90 mm and a length of 500 mm is presented by Fig. 13.2, which is like that of Ref. [24]. The stainless wire-mesh screen with a diameter (d_W) of 0.36 mm was employed as porous media, and it was provided as six bundles to attach inside the pipe. The distance between two adjacent porous bundles (L) was kept at 80 mm. In the experiment, there were two interesting parameters: porosity (ε) converted from PPI, and porous thickness (H). Table 13.1 summarizes the 25 cases, constructed from five ε and five H, for measuring the pressure drop (ΔP) and evaluating K and F.

As shown in Fig. 13.1, an experimental procedure was the same as in a previous work [24]. To get the understanding again, however, the experimental procedure was represented. In the beginning, the fresh air at the ambient condition (i.e., inlet air temperature, T_{in}, of 27 °C) was supplied by an air blower into the system. Then, the airflow rate (Q) controlled with a valve was monitored by Rotameter.

The Q was changing at the range of 0.00167 m^3/s (100 LPM)–0.01167 m^3/s (700 LPM). The fresh air flowed past the setting pipe to provide a uniform streamline

Table 13.1 Testing cases of wire-mesh porous media

Case	PPI	Porosity (ε)	H (mm)
1	6	0.866	1.44 (4 screen)
2	6	0.866	2.88 (8 screen)
3	6	0.866	4.32 (12 screen)
4	6	0.866	5.76 (16 screen)
5	6	0.866	7.20 (20 screen)
6	8	0.822	1.44 (4 screen)
7	8	0.822	2.88 (8 screen)
8	8	0.822	4.32 (12 screen)
9	8	0.822	5.76 (16 screen)
10	8	0.822	7.20 (20 screen)
11	10	0.777	1.44 (4 screen)
12	10	0.777	2.88 (8 screen)
13	10	0.777	4.32 (12 screen)
14	10	0.777	5.76 (16 screen)
15	10	0.777	7.20 (20 screen)
16	12	0.733	1.44 (4 screen)
17	12	0.733	2.88 (8 screen)
18	12	0.733	4.32 (12 screen)
19	12	0.733	5.76 (16 screen)
20	12	0.733	7.20 (20 screen)
21	14	0.688	1.44 (4 screen)
22	14	0.688	2.88 (8 screen)
23	14	0.688	4.32 (12 screen)
24	14	0.688	5.76 (16 screen)
25	14	0.688	7.20 (20 screen)

and the uniform flow of air was completely obtained while passing the straightened pipe. In addition, to generate the full developing flow in a pipe, its length of 900 mm (10D) [29] was aligned horizontally in front of a testing pipe. Finally, the pressure drop (ΔP) across the testing pipe was measured by a digital manometer with an accuracy of $\pm 0.3\%$.

2.2 Estimated Equation of Permeability (K) and Inertia Coefficient (F)

In evaluation of permeability (K) from the data accumulated by the experimental procedure, the Forcheimer equation [9] or Eq. (1) can be calculated by:

$$\Delta P = au + bu^2 \tag{2}$$

$$a = \frac{\mu L_0}{K} \tag{3}$$

$$b = \frac{\rho F L_0}{\sqrt{K}} \tag{4}$$

Here, u is air velocity with the range of 0.262 m/s–1.833 m/s, which is calculated by $u = Q/A$ (A is the cross-section area of a testing pipe). For the value of L_0, H is applied.

Based on Ref. [24], Eq. (2) was transformed into a linear equation as follows:

$$\frac{\Delta P}{u} = a + bu \tag{5}$$

where variable a is the slope of the line and b is the y-axis ($\Delta P/u$) intercept. Eq. (5) was fitted by the linear regression method [30]. Such two variables (a and b) were easiest to obtained. Thus, after K was estimated by $\mu L_0/a$, F was automatically given by $bK^{1/2}/\rho L_0$.

3 Results and Discussion

3.1 *Effect of* ε *on* K *and* F

Figure 13.3a presents the relationship between the ratio of pressure drop to air velocity ($\Delta P/u$), and air velocity (u) for the effect of porosity (ε) at the experimental condition of porous thickness (H) of 7.2 mm. From the experiment, it was found that the $\Delta P/u$ was increased as rising u owing to the nature of fluid flow in the pipe [24, 31, 32]. As a constant of u, the pressure drop ($\Delta P/u$) was reduced as increasing ε (PPI reduce) because of the difficult flow that occurred at a lower porosity (ε) [32–35]. In the figure, the solid lines indicate the fitted curves across the experimental points (symbols) based on regression analysis. The coefficients of determination (R^2) between the solid line and the symbols were all more than 0.98, indicating that this regression was credible [30].

In Table 13.2, the solid lines (linear equations) are summarized. The coefficients a and b specified in Eq. (5) may be found by fitting a linear equation across these points. By substituting the a and b from this figure into Eqs. (5) and (6), the K and the F were obtained. The relationship of K and F on ε at the $H = 7.2$ mm is illustrated in Fig. 13.3b. As ε increased, the K was decreased but the F grew. This was described by the effect of a raiser pressure drop (ΔP) similar with the result in Fig. 13.3a. The trend of K and F of the present study was consistent to the available works [28, 36, 37].

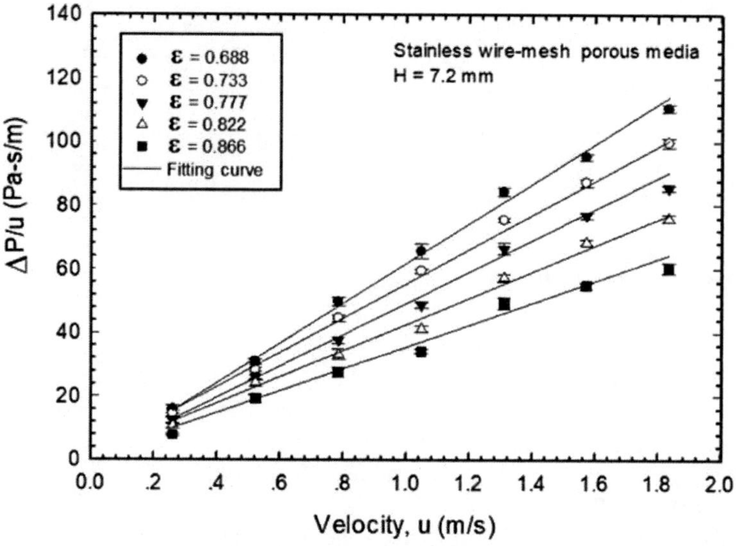

(a) relation between ΔP/u and u

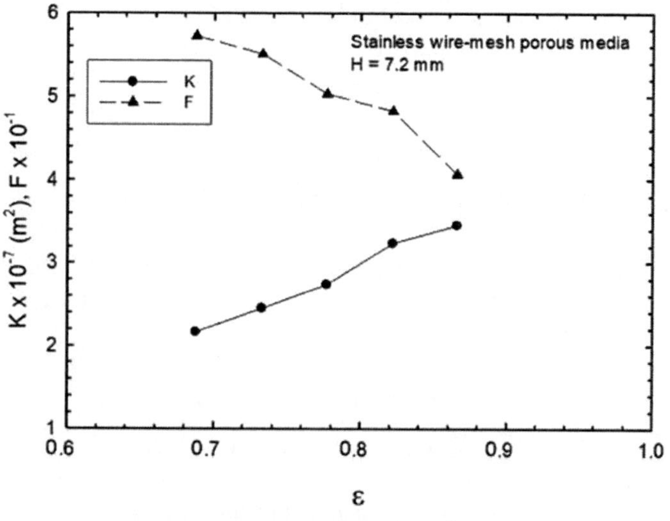

(b) relation of K and F on ε

Fig. 13.3 Effect of ε on $\Delta P/u$, K and F. (**a**) Relationship between $\Delta P/u$ and u. (**b**) Relationship of K and F on ε

Table 13.2 Linear equation of $\Delta P/u$ for effect of ε

Porosity (ε)	Linear equation	R^2
0.866 (PPI = 6)	$\Delta P/u = 0.386 + 34.142u$	0.986
0.822 (PPI = 8)	$\Delta P/u = 0.411 + 41.996u$	0.993
0.777 (PPI = 10)	$\Delta P/u = 0.486 + 47.439u$	0.995
0.733 (PPI = 12)	$\Delta P/u = 0.543 + 55.252u$	0.997
0.688 (PPI = 14)	$\Delta P/u = 0.614 + 61.086u$	0.996

3.2 *Effect of* H *on* K *and* F

Figure 13.4a presents the relationship between the $\Delta P/u$ and u at the porosity (ε) of 0.688 (PPI = 14) for the effect of porous thickness (H). The $\Delta P/u$ was increased with u, which was the same trend in effect for ε (see Fig. 13.3a). For constant of u, the trend of pressure drop ($\Delta P/u$) was raised as increasing H because it was the effect of a thicker porous media, resulting in a more difficult flow of air through the porous media had occurred [24, 31, 32].

In considering the fitting curves, the solid lines gave good agreement to the experiments (symbols). The determination coefficients (R^2) of the developed equations became higher than 0.96, and they were summarized in Table 13.3. Figure 13.4b illustrates the relationship between the K and the F on the impact of H. As H increases, the trend of K was decreased, but the F was rose. This can be explained by the same reason shown in Fig. 13.4a.

3.3 *Regression Analysis of* K *and* F *to* ε *and* H

From the 25 cases of the present experiment based on the range of $0.733 \leq \varepsilon \leq 0.866$ and $1.44 \leq H \leq 7.20$ mm, two equations of K and F generated by the multiregression method are given as follows:

$$K = (1.3425 + 6.2161\varepsilon - 0.4977H) \times 10^{-7} \quad (6)$$

$$F = 0.6296 - 0.7524\varepsilon + 0.0653H \quad (7)$$

To validate both equations, the comparison results of equations (Eqs. (6) and (7)) to available empirical equations are reported by Fig. 13.5. The equation proposed by Gebart [21] and Koponen et al. [23] were employed to validate the K (see Fig. 13.5a), while the validation of F was compared with Du Plessis's equation alone (see Fig. 13.5b) [22].

As seen in Fig. 13.5a, the solutions from the present equation of K at the H = 7.20 mm gave good agreement with both available works [21, 23]. For

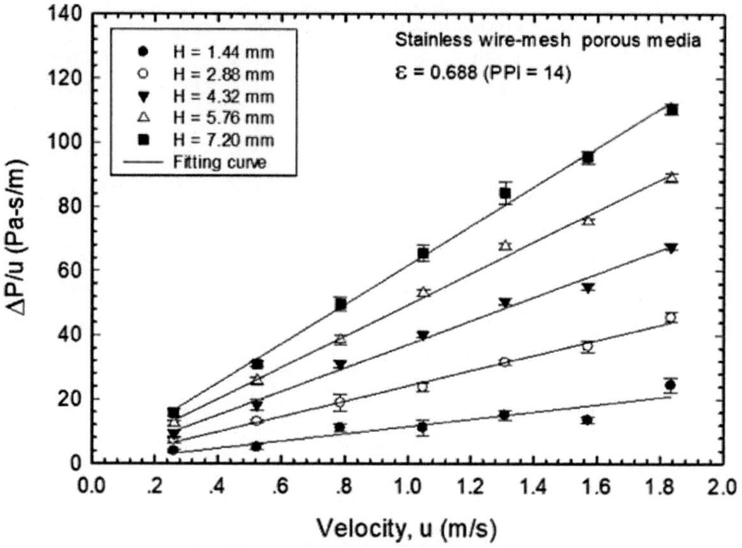

(a) relation between ΔP/u and u

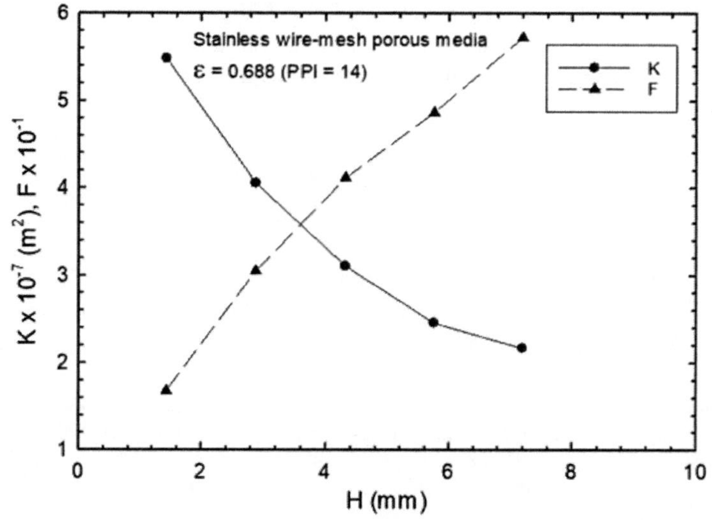

(b) relation of K and F on H

Fig. 13.4 Effect of H on $\Delta P/u$, K and F. (**a**) Relationship between $\Delta P/u$ and u. (**b**) Relationship of K and F on H

Table 13.3 Linear equation of $\Delta P/u$ for effect of H

Porous thickness (H)	Linear equation	R^2
1.44 mm (4 screen)	$\Delta P/u = 0.243 + 11.288u$	0.965
2.88 mm (8 screen)	$\Delta P/u = 0.328 + 23.802u$	0.994
4.32 mm (12 screen)	$\Delta P/u = 0.428 + 36.657u$	0.992
5.76 mm (16 screen)	$\Delta P/u = 0.543 + 48.838u$	0.996
7.20 mm (20 screen)	$\Delta P/u = 0.614 + 64.086u$	0.996

comparison of F at $H = 7.20$ mm, as shown in Fig. 13.5b, the same trend of the previous equation and Du Plessis' s model [22] was achieved. A higher level from that equation, however, appeared. Thus, it can be said that the accuracy of the proposed correlation was practical use in the examined range.

4 Conclusion

An experimental and analytical study investigating the permeability (K) and the inertia coefficient (F) of wire-mesh porous media, based on Forchheimer principle, was proposed. Stainless SUS304 with different porosity (ε or PPI) and thicknesses (H) were examined. The trend of K increased and F decreased when ε was raised. The opposite result was obtained by increasing H: K decreased and F increased. From the fitting process, equation of K and F, estimated by multi-regression method, can be shown by

$$K = (1.3425 + 6.2161\varepsilon - 0.4977H) \times 10^{-7} \text{ and}$$
$$F = 0.6296 - 0.7524\varepsilon + 0.0653H$$

For which the accuracy was in the range of $0.733 \leq \varepsilon \leq 0.866$ and $1.44 \leq H \leq 7.20$ mm. Good agreement between this chapter's equations and other available works [21–23] was obtained.

Acknowledgments The authors sincerely appreciate the finance support from Faculty of Engineering and Architecture, Rajamangala University of Technology Isan (RMUTI) for the registration fee. Thanks also to Mr. Bunpot Boorana, Mr. Kittichai Muangkoed, and Mr. Julladid Suksabai, undergraduate students of the Mechanical Engineering Department, for recording the experimental data.

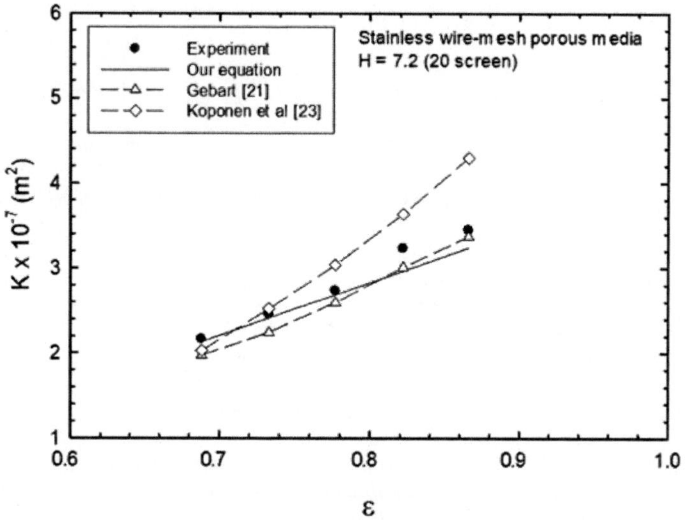

(a) relation between K and ε

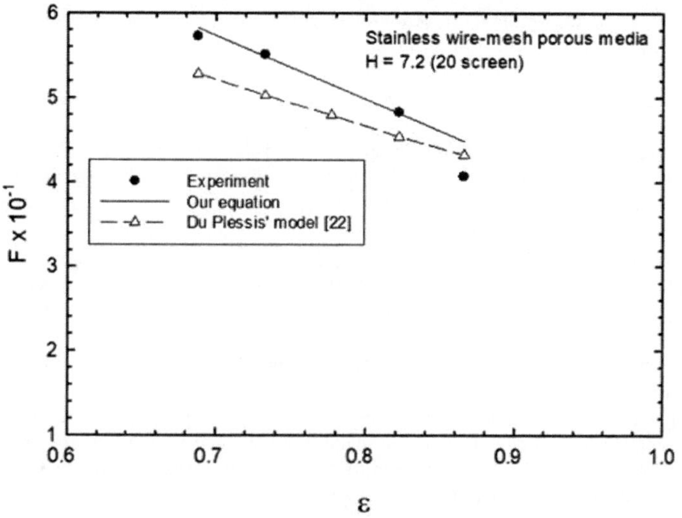

(b) relation between F and ε

Fig. 13.5 Comparison of K and F between the present equation with available works. (**a**) Relationship between K and ε. (**b**) Relationship between F and ε

References

1. M.E. Beygi, F. Rashidi, Analytical solutions to gas flow problems in low permeability porous media. Transp. Porous Media **87**, 421–436 (2011)
2. K. Arash, H.M. Mehdi, A.R. Deresh, Reliable method for the determination of surfactant retention in porous media during chemical flooding oil recovery. Fuel **158**, 122–128 (2015)
3. A. Telmadrreie, D. Ankit, J.T. Japan, K. Ergun, C. Phillip, CO_2 microbubbles-a potential fluid for enhanced oil recovery: Bulk and media studies. J. Pet. Sci. Eng. **138**, 160–173 (2016)
4. D. Bernard, W.M. John, Groundwater age in fractured porous media: Analytical solution for parallel fractures. Adv. Water Resour. **37**, 127–135 (2012)
5. F. Alessio, S. Anna, A numerical method for two-phase flow in fractured porous media with non-matching grids. Adv. Water Resour. **62**, 454–464 (2013)
6. F. Ali, L. Faical, Two-phase flow hydrodynamic study in micro-packed beds – Effect of bed geometry and particle size. Chem. Eng. Process. **78**, 27–36 (2014)
7. T.F. Zhao, C.Q. Chen, The shear properties and deformation mechanisms of porous metal fiber sintered sheets. Mech. Mater. **70**, 33–40 (2014)
8. K.N. Kim, J.H. Kang, S.G. Lee, J.H. Nam, C.J. Kim, Lattice Boltzmann simulation of liquid water transport in microporous and gas diffusion layers of polymer electrolyte membrane fuel cells. J. Power Sources **278**, 703–717 (2015)
9. P. Forchheimer, Wasserbewegung durch boden. VDI Z. **45**, 1782–1788 (1901)
10. S. Ergun, Fluid flow through packed column. Chem. Eng. Sci. **48**, 89–94 (1952)
11. H. Darcy, *Les Fontaines Publiques de la ville de Dijon* (Dalmont, Paris, France, 1856)
12. A. Nabovati, E.W. Llewellin, C.M. Antonio, A general model for the permeability of fibrous porous media based on fluid flow simulations using the lattice Boltzmann method. Compos. Part A **40**, 860–869 (2009)
13. J. Kozeny, Ueber kapillare Leitung des Wassers in Boden. Stizungsber Akad Wiss Wien. **136**, 271–306 (1927)
14. P.C. Carman, Permeability of saturated sands, soils and clays. J. Agric. Sci. **18**, 262–273 (1939)
15. K. Vafai, C.L. Tien, Boundary and inertia effects on convective mass transfer in porous media. Int. J. Heat Mass Transf. **25**(8), 1183–1190 (1982)
16. Joseph, D.A. Nield, G. Papanicolaou, Nonlinear equation governing flow in a saturated porous medium. Water Resour. Res. **18**, 1049–1052 (1982)
17. B. Antohe, J.L. Lage, D.C. Price, R.M. Weber, Experimental determination of permeability and inertial coefficient of mechanically compressed aluminium metal layers. ASME J. Fluids Eng. **119**, 404–412 (1997)
18. M.D. Aminu, B.U. Ardo, A novel approach for determining permeability in porous media. J. Pet. Environ. Biotechnol. **6**, 1–5 (1994)
19. A. Eshghinejadfard, L. Daroczy, G. Janiga, D. Thevenin, Calculation of the permeability in porous media using the lattice Boltzmann method. Int. J. Heat Fluid Flow **62**, 93–103 (2016)
20. P. Yang, Z. Wen, R. Duo, X. Lui, Permeability in multi-size structures of random packed porous media using three-dimensional lattice Boltzmann method. Int. J. Heat Mass Transf. **106**, 1368–1375 (2017)
21. B.R. Gebart, Permeability of unidirectional reinforcements for RTM. J. Compos. Mater. **26**(8), 1100–1133 (1992)
22. P. Du Plessis, A. Montillet, J. Comiti, J. Legrand, Pressure drop prediction for flow through high porosity metallic foam. Chem. Eng. Sci. **49**, 3545–3553 (1994)
23. A. Koponen, D. Kandhai, E. Hellen, Permeability of three-dimensional random fiber webs. Phys. Rev. Lett. **80**, 716–719 (1998)
24. N. Huchaiyaphum, S. Panpanya, B. Krittacom, Permeability of wire-net porous media determined by a simple Darcy-Forchheimer equation. J. Res. Appl. Mech. Eng. **6**(1), 63–71 (2018)

25. P. Waramit, P. Chanmak, R. Peamsuwan, B. Krittacom, Forced convection enhancement of air flowing inside circular pipe with varying the pitch (P) of wire-mesh porous media. Energy Rep. **7**(sup. 5), 70–82 (2021)
26. B. Krittacom, and A. Klamnoi: A simplified linear scattering phase function for solving equation of radiative heat transfer, presented in Proceedings of the 4th International Symposium on Engineering, Energy and Environments, Pattaya, Thailand, November 8–10 (2015), pp. 457–463.
27. A.A. Mohamad, Heat transfer enhancements in heat exchangers fitted with porous media part I: Constant wall temperature. Int. J. Therm. Sci. **42**(4), 385–395 (2003)
28. B.I. Pavel, A.A. Mohamad, An experimental and numerical study on heat transfer enhancement for gas heat exchangers fitted with porous media. Int. J. Heat Mass Transf. **47**(23), 4939–4952 (2004)
29. Y.A. Cengel, A.F. Ghajar, *Heat and Mass Transfer: Fundamental and Applications*, 4th edn.9 (McGraw-Hill Inc., Singapore, 2011)
30. W. Cheney, D. Kincaid, *Numerical Mathematics and Computing*, 7th edn. (Brooks/Cole, Cengage Learning, Boston, USA, (2013)
31. L. Green Jr., Fluid flow though porous metals. J. Appl. Mech. **18**, 39–45 (1951)
32. M. Kaiany, *Principle of Heat Transfer in Porous Media*, 2nd edn. (Springer-Verlag, New York, 1996)
33. B. Krittacom, P. Amatachaya, R. Sangchot, Energy balance in Al-Co open-celled foam of transpiration cooling. Appl. Mech. Mater. **575**, 41–45 (2014)
34. P. Amatachaya, B. Krittacom, Combustion mechanism of gas porous burner installed an in-line tube-bank heat exchanger. Energy Procedia **138**, 50–55 (2017)
35. P. Waramit, P. Chanmak, R. Peamsuwan, B. Krittacom, Forced convection enhancement of air flowing inside circular pipe with varying the pitch (P) of wire-mesh porous media. Energy Rep. **7**(Supple 5), 70–82 (2021)
36. A. Bhattacharya, V.V. Calmidi, R.L. Mahajan, Thermophysical properties of high porosity metal foam. Int. J. Heat Mass Transf. **45**, 1017–1031 (2002)
37. H. Cho, N. Jeong, H.J. Sung, Permeability of microscale fibrous porous media using the lattice Boltzmann method. Int. J. Heat Fluid Flow **44**, 435–443 (2013)

Index

A
Alpha wave, 30
ARP cache poisoning, 155–168
ARP protocol, 156–158
ARP spoofing attack, 158, 159, 163, 164
Attention, 30, 34–37, 172
Automatic Face Recognition (AFR), 13, 14, 17

B
Beta wave, 30, 37
Brain-machine interface (BMI), 30, 37

C
Case study, 78–87
Cash register, 90–92, 94–98, 100
Classification, 6, 34, 41, 43, 45, 55, 160
Colour calibration, 104–106, 114
Comparison, 25, 45, 49, 54, 57, 91, 105, 110, 129–131, 160, 173, 178
Convolutional neural network (CNN), 54, 55
Covariance matrix, 14, 15, 17, 20
Crowdsourcing, 103–115
Cultural heritage conservation, 103–115
Customers' burstiness, 91, 92, 100
Customers' waiting time, 90

D
Darcy-Forchheimer equation, 171–181
Deep learning, 4, 6, 54, 78–87, 105
Detection of gaps, 77–87

E
Eigen vector, 14, 15, 17–19, 22
Electroencephalography (EEG), 30, 32–35
Electromyography (EMG), 32, 35, 40–45, 48–51
Enterprise architecture debt, 134–150
Enterprise architecture debt management, 134–136
Enterprise architecture management, 135, 136

F
Face detection, 53–61
Faster-RCNN, 4, 6–9
Feature vector, 14, 17, 18, 23
Frequent patterns, 117, 118, 120

G
Game control, 40, 44
Gap distribution function, 92–97
Generating functions, 90–100

I
ICMP protocol, 157, 161, 162, 166, 168
Image processing, 85, 115
Itemsets mining, 128

L
Linear discriminate analysis (LDA), 14, 19–26

C.-C. Wang, A. Nallanathan (eds.), *Proceedings of the 5th International Conference on Signal Processing and Information Communications*, Signals and Communication Technology, https://doi.org/10.1007/978-3-031-13181-3

M
Machine learning, 40, 78–80, 84, 87, 159, 160
Man-in-the-middle (MitM), 156, 166, 168
Minimum Euclidian distance, 18
MTCNN, 53–57, 59–61
Multi-camera switching, 3–10

O
Object detection, 6, 54
Objective indicators, 72

P
Periodic-frequent pattern, 117–131
Periodic patterns, 117–131
Permeability, 172, 173, 175–176, 180
Principle component analysis (PCA), 13–26

R
Rate of recognition (ROC), 19
Recreation space, 64, 65, 69, 73, 74
Rehabilitation, 40, 41

S
Sheet metal industry, 78–87
Sound field simulation, 71
Soundscape perception, 65–67, 71–73

T
Transactional database, 118, 121, 123, 131

V
Video and camera tracking, 4–6, 9
Viewpoint, 134–136, 138–152
Viola-Jones, 53–55, 57–61
Visual reality, 39–51

W
Wire-mesh porous media, 173, 175, 180